Coal

The Energy Source

Coal

The Energy Source of the Past and Future

Harold H. Schobert

AMERICAN CHEMICAL SOCIETY
WASHINGTON, DC 1987

Library of Congress Cataloging-in-Publication Data

Schobert, Harold H., 1943–
Coal, the energy source of the past and future.

Bibliography: p.
Includes index.

1. Coal. I. Title.

TP325.S34 1987 662.6′2 87-11433
ISBN 0-8412-1171-X
ISBN 0-8412-1172-8 (pbk.)

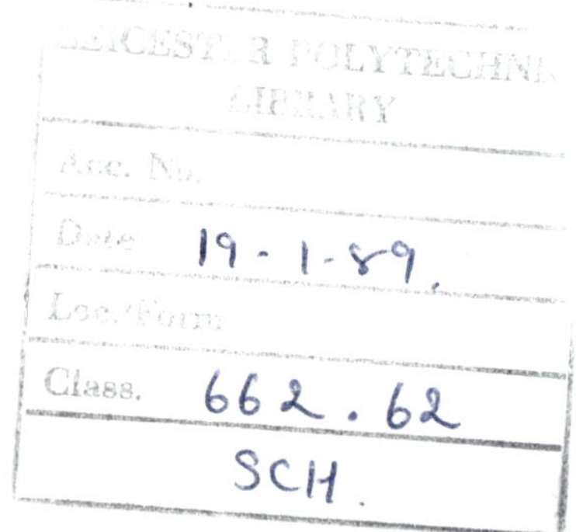

PRINTED IN THE UNITED STATES OF AMERICA

About the Author

HAROLD H. SCHOBERT is an associate professor in the Fuel Science Program of the Department of Materials Science and Engineering at The Pennsylvania State University. He received a B.S. in chemistry from Bucknell University and a Ph.D. from Iowa State. He is a coauthor of about 50 papers, government reports, and book chapters on various aspects of coal research. His current research interests include the behavior of coal ash at high temperatures in combustion and gasification systems, low-severity coal liquefaction, and the geochemistry of lignites. Schobert is the editor of a previous ACS book, *The Chemistry of Low-Rank Coals.*

Contents

Preface

Our way of life depends on the convenient and easy use of energy: Some form of energy provides heat and light for our homes. Any time we drive a car or take public transportation, we use fuel. Fuels provide the energy for our factories and farms. Virtually every item that we buy or use comes from the chemical industry or the metals industry. These industries use energy to transform raw materials into the finished products of commerce.

For most of us, energy is amazingly accessible: flicking a light switch, turning the ignition of a car, cooking dinner, or riding in a multiengine jet plane. Energy is so available that it intrudes on our consciousness only when it is big news. In the 1970s, the Organization of Petroleum Exporting Countries imposed an oil embargo that caused stunning increases in the price of crude oil and resulted in turmoil. What we called an energy crisis in those days had significant effects on how we lived. Many of us bought smaller, more fuel-efficient cars or looked for ways to add more insulation to our homes. In the mid-1980s, oil prices began tumbling to levels that were one-third or less of what they were before the price collapse. Suddenly, the turmoil seemed to end, except for the people working in the oil industry. Then, in 1986, the nuclear reactor meltdown at Chernobyl in the Soviet Union focused the world's attention on another aspect of energy technology.

Coal was undoubtedly our most important energy resource between the Civil War and World War II. Long-distance travel and shipment of freight depended on railroads serviced by coal-fired steam locomotives and on steamships heated and powered by coal-fueled boilers. Coal furnaces heated homes, and gas made from coal was used for cooking, heating, and lighting. Coal also fueled factories, and coke, which is a product made from coal, smelted iron and steel. Coal byproducts were the foundation of the chemical industry and gave us plastics, dyes, solvents, medicines, and fertilizers.

During the energy crisis in the 1970s, attention was focused again on coal. People realized during this time that coal was an important energy source for the United States. Lately, however, the price of oil has dropped. This situation has reduced concern about energy sources and has turned attention away from coal. Nevertheless, despite the fluctuations in the oil economy, we are producing and consuming oil faster than we are finding it. Inevitably, supplies will someday be scarce. For the future, we need to develop an economy based on renewable or inexhaustible forms of energy that are nonpolluting. Meanwhile, we need a stable, dependable source of energy that will sustain us from the decline of the oil economy to the advent of widespread commercial technology based on new energy sources. Because of the aversion to nuclear power in the United States and the recent Soviet catastrophe, we have only one option: coal.

The development of new technologies for using coal requires that we know something about the substance. The most comprehensive book on the science and technology of coal is the second supplemental volume of *Chemistry of Coal Utilization,* which is 2229 pages long. I have not tried to duplicate this effort in my book. Instead, I have concentrated on telling the story of coal to the general public in an attempt to convey the important issues and to explain how coal is formed, where it is found in the world, how scientists study it, how it is mined, the ways we use it, and how we can use it in the future. Because this book is much smaller than *Chemistry of Coal Utilization,* much of the total body of knowledge has been left out or glossed over. Suggested sources for probing further into the story of coal are listed at the end of this book; most of these sources can be found in libraries.

During initial discussions of this project, the American Chemical Society suggested that the book be written for the "intelligent lay person". I suspect that the intelligent lay person is as elusive a character as the mythical "average person". I have assumed that most people reading this book will have had a course in chemistry, probably the introductory college course, as well as some exposure to the principles or terms of other sciences through introductory courses in college or high school. Nevertheless, I feel that people who have had almost no science instruction beyond the high school level can read most of this book with profit and, I hope, enjoyment. This book is not intended to be a professional monograph on coal, but I trust that some of my colleagues will find it useful.

Acknowledgments

Most of the work on this book was done while I was employed at the University of North Dakota Energy Research Center (UNDERC). I am grateful to George Wiltsee, the director of UNDERC, who made it possible for me to have the drafts of this book typed by the UNDERC word processing group and to obtain help from the graphics staff for the illustrations. The typing of the manuscript was done by Rosemary Honek and her staff; Paul Gronhovd provided expert help with the graphics. My administrative assistant, the inimitable Nita Ralston, skillfully nudged the manuscript drafts and rough drawings through the bureaucratic labyrinth into final form.

At one time or another virtually every member of the former Coal Science Division of UNDERC helped by answering my questions or providing advice, specific facts, or illustrations. Other colleagues at UNDERC, the University of North Dakota, and the Grand Forks Project Office of the U.S. Department of Energy also provided information. William Spackman, Professor Emeritus of Paleobotany at The Pennsylvania State University, reviewed the final draft and provided many helpful comments. I also received help and advice from Erle Diehl, recently retired from Bituminous Coal Research, who read early drafts of various chapters and the entire final draft. Bonnie Grove of The Pennsylvania State University typed the final manuscript revisions and the figure captions.

Wayne Kube deserves special mention. In the early stages of organizing and writing this book, Wayne enthusiastically supported my efforts. He was then in the process of retiring from a long and illustrious career in the Chemical Engineering Department at the University of North Dakota and very generously gave me a sizable portion of his personal professional library for reference material. I regret that Wayne did not live to see the final product.

This book was written entirely on my own time and, except for the invaluable help that I have acknowledged here, entirely with my own resources. Therefore, any opinions or prejudices that appear in this book are not necessarily those of the University of North Dakota or The Pennsylvania State University. They are entirely my own, and so are the mistakes.

HAROLD H. SCHOBERT
January 16, 1987

Chapter 1

The Substance Called Coal

What is coal? The answer to this question depends on whom you ask. Coal is a vital national resource that could replace imported energy supplies. Coal is the key to a smooth transition from our present petroleum-based energy economy to renewable, nonpolluting energy sources in the 21st century. Coal is an obsolete fuel whose predominance as an energy source passed decades ago. Coal is a potential source of clean-burning gas or liquid fuels. Coal is a dirty fuel; its mining and use have caused environmental damage as well as human misery and suffering. Coal is a fascinating, complex, scientific jigsaw puzzle; the more pieces we put into place, the more we learn about our earth's history. Coal is an uninteresting black rock; because all we're going to do is dig it up and burn it, we should not be very concerned about its composition or how it got to be that way.

Which of these answers is correct? Each has a measure of truth. The United States has abundant reserves of coal. If we wished to accept the economic and industrial dislocation that might occur, coal or gases and liquids made from coal could be used in place of the petroleum and natural gas we now import. Eventually, supplies of petroleum and gas, imported or from domestic sources, will begin to decline. Coal or coal-derived substitute fuels can take up the slack until new technologies become economically successful. Although the role of coal in the transition to some future economy based on new energy sources appears very promising, most likely the conventional ways of using coal are past their prime. New, more efficient ways of burning coal will make present power-plant technology obsolete. The last new plant in the United States to make metallurgical coke from coal has probably been built already.

Technologies are available to reduce or eliminate the pollution caused by burning coal. Labor laws and health and safety regulations have greatly improved the working conditions of coal miners. Still, we cannot rewrite history and pretend that coal was never a dirty, smoky fuel and that coal mines were not virtual dungeons. To some, perhaps, coal is indeed just the black matter that has to be shoveled in one end of a plant so that electricity, gas, steel, or steam comes out the other end. But to those who have devoted their careers to understanding the chemical constitution and physical properties of coal, to working out its geochemical history, or to learning how and why it reacts, coal chemistry is an exciting intellectual challenge.

The value of coal is illustrated by considering the needs of a modern industrial civilization:

- electricity;
- transportation fuels;
- metals and the chemicals for extracting the metals from their ores;
- chemicals, including plastics, medicines, and fertilizers; and
- process heat, that is, the heat used for the boiling, melting, annealing, and shaping transformations in the chemical and metal industries.

Coal can help satisfy all these needs. Today, about one-half of our electricity is generated in power plants that burn coal.[1] Coal was used as a transportation fuel in the days of steam locomotives and coal-fueled steamships. Those days aren't quite over yet. Steam locomotives are still the workhorses of railroads in China, and radical new designs of coal-fired locomotives are being considered for service in North America. Liquid fuels produced from coal are the chief means of support of South African transport. During World War II, coal-derived liquid fuels were manufactured and used in Germany. Metallurgical coke made from coal is the agent used to liberate iron from its ores. Except for the ore itself, coke is the most important raw material in the iron and steel industry. Today, coke production is the second most important use of coal.[2] Traditionally, coal itself has not been considered a source of metals, but in recent years the ash remaining after coal is burned has been studied as a possible source of aluminum and rare metals such as germanium needed by the electronics industry. For many years, the byproducts from coke manufacture were a source of a tremendous variety of chemicals, including solvents, fertilizers, drugs, dyes, perfumes, and artificial flavors. Most of our modern organic chemical industry is based on petroleum rather than coal, but when petroleum eventually becomes less plentiful, a new chemical industry, deriving not from coke byproducts but from newer methods of converting coal, will take its place. Coal can be burned to provide heat for chemical or metal processing, or it can be used to generate steam for a heat or energy source.

Coal Through the Years

Nobody knows who first used coal or even where it was first used. Deposits of coal are exposed on the earth's surface in many places around the world. Someone eventually noticed that this black rock would burn and could provide warmth and heat for cooking. Such a discovery was probably made independently in several places in the world.

Ancient World

The first large-scale use of coal began in China thousands of years ago. Coal was used in the metallurgical industry and as a source of domestic heat as long ago as A.D. 300. By A.D. 1000, coal was the predominant fuel and energy source in China.[3] Knowledge of the Chinese use of coal was first brought to the West by the Venetian traveler Marco Polo, who spent 24 years in China. When Marco Polo returned to Venice in 1295, he reported the digging of a black stone that was burned as a fuel.

In Western civilization the first person who might have laid claim to being a coal researcher was Theophrastus. He was a pupil of Aristotle who succeeded his master as the head of the Lyceum in Athens in 323 B.C.[4] All of Theophrastus' work, which included studies of zoology, physics, and botany, was characterized by a drive to seek and strengthen the unifying principles and facts of those subjects. Around 300 B.C., Theophrastus published *De Lapidus,* which remained the best reference book on stones and minerals for the next 2000 years. Theophrastus described a black stone that would burn and was sometimes used by blacksmiths instead of charcoal. He called this burning stone *anthrax;* this word is the root of our word *anthracite,* a type of coal.[5]

When the Roman legions marched north, they found coal-mining operations in progress near what is now St. Etienne in France. They also found coal being used in Great Britain and adopted the practice for themselves.

Middle Ages

During the Middle Ages, coal was mined in Germany as early as the 900s.[6] In France, coal was mentioned in the charter of the priory of St. Sauveuren-

Rue in a document dated 1095. One of the earliest medieval treatises to discuss coal was published about a hundred years later and was written by Reinier, a monk in Liège. Reinier wrote about a black "earth" that behaved like charcoal and was used by metalworkers. During the 13th century, commercial coal mining became increasingly widespread throughout Great Britain and continental Europe. Liège in Belgium, Lyonnais in France, and Newcastle in England became important mining centers. Internal coal trade was established in Great Britain at this time, and export business began with English coal being shipped to Brugge. Two-way traffic began about a century later. In 1328, ships carried wheat from France to England and loads of English coal back to France.[7]

Until this time the principal uses of coal had been for domestic heating and as an occasional substitute for charcoal in metalworking. (Despite its name, charcoal is made from wood.) In the 13th century, coal began to be exploited as a source of process heat as well. In 1226, a street in London known as Lime Burners Lane also became known as Sea Coal Lane.[8] Lime was used in early industry as an ingredient of glass, mortar, and whitewash and as a chemical in the tanning of hides. It was made by heating or, in crude terminology, burning limestone in kilns. Sea coal became a useful fuel for firing lime kilns. By the 14th century, the use of coal became widespread enough that mining in Great Britain had spread through the Midlands, Derbyshire, Shropshire, and Nottinghamshire and into Wales and Scotland.

Unfortunately, the dark side of the coal industry accompanied this expansion of coal mining and use. Most of the mining in the medieval period was done by digging open pits to get at the coal. So much mining activity was going on around Newcastle that it actually became dangerous to approach the city in the dark because of the likelihood of falling into a coal pit. In 1243, Ralph Ulger drowned in a coal pit and thereby attained the unfortunate distinction of being the first recorded fatality ascribable to coal mining—the first of untold thousands to come.

Sea coal and other coals mined near the earth's surface burned with very smoky flames and fumes of unburned material. Coal had already developed a reputation as a dirty fuel, and concerns were expressed about the consequences of increased coal use. The first person of prominence to suffer the effects of pollution from coal burning was Eleanor of Castile, the wife and queen consort of Edward I of England. Eleanor is perhaps best known as the heroine of the romantic, if highly improbable, tale of having saved Edward's life during a Crusade by sucking the poison from a dagger wound he suffered in battle near Acre in the Holy Land. Her brief appearance in the story of coal came in 1257 when she abandoned Nottingham Castle, driven out by the fumes from the sea-coal fires in the castle.[9]

London holds the dubious honor of being the first city to experience severe air pollution. Complaints were raised in 1285 and again in 1288 about the deterioration of air quality resulting from the extensive use of coal for firing lime kilns. The complaints eventually resulted in the issuance of a royal proclamation forbidding the use of sea coal because an "intolerable smell diffuses itself throughout the neighboring palaces and the air is greatly affected to the annoyance of the magnates, citizens, and others there dwelling and to the injury of their bodily health."[10] Some use of coal continued in spite of the prohibition. A few years later, a commission was appointed to investigate the burning of sea coal in London. This commission had the power to fine a first-time offender and actually demolish the offender's facilities or equipment upon a second offense.[10]

Renaissance

Proclamations and commissions notwithstanding, the use of coal continued. In 1578, Elizabeth I complained of the smoke from coal fires used in breweries. The London Company of Brewers agreed, apparently voluntarily, to use wood as the fuel in breweries near Westminster for the comfort of the queen.[11] During the Elizabethan era, fastidious people avoided entering dwellings or rooms in which coal fires were burning. Not everyone was so picky. One of Shakespeare's characters, Mistress Quickly, the hostess of the Boar's Head Tavern, tells Falstaff,[12] "Thou didst swear to me upon a parcel-gilt goblet, sitting in my Dolphin Chamber, at the round table, by a sea-coal fire, upon Wednesday in Whitsam week."

Coal mining on a large scale began in Great Britain in the mid-16th century, about the time Elizabeth ascended the throne.[13] Wood or charcoal made from wood were still the preferred fuels, but in the early 17th century, severe shortages of wood began to be felt in Great Britain and on the Continent. More countries increased their coal production to make up for the decreasing availability of wood, and factories converted to coal as their primary fuel.

Industrial Revolution

The conversion from wood to coal as the predominant energy source during the next hundred years led to two major discoveries. As coal mines were dug deeper into the earth, a power source was needed to operate water pumps to keep the mines dry. This need led to the development of the steam engine. As charcoal became increasingly scarce, a method was needed to substitute coal for charcoal in iron-ore smelting. Iron makers learned how to convert coal to coke, a development that made possible the large-scale production of inexpensive iron. Thus, the steam engine and cheap iron provided the foundation for a technology so extensive and so remarkable that it became a crucial turning point in the history of Western civilization: the Industrial Revolution.

One of the significant changes brought by the Industrial Revolution was a wholesale shift from small-scale manufacturing or work at home to large factories. The operation of a large factory required a reliable source of power, a job for which the steam engine was admirably suited. The widespread use of steam engines on a large scale represented in turn a significant demand for large quantities of inexpensive fuel. The only energy supply then available that could meet this need was coal.

The Industrial Revolution began in England and spread during the 19th century to other parts of Europe and North America. The countries best able to accommodate the new economic changes wrought by the Industrial Revolution were those having abundant supplies of coal, principally, the United States, Germany, and France. To some extent the political strength of countries in the 19th century derived from possessing or having access to coal. In those days coal meant power, power not only to drive machinery or to produce electricity, but geopolitical power as well. Otto von Bismarck, the architect of German foreign policy for almost three decades in the late 19th century, stated, "It is not by speeches and resolutions that the great questions of the time are decided, but by blood and iron."[14] Bismarck's "iron" was the iron of guns and bayonets. This "Blut und Eisen" speech, made in 1862, indeed set the tone for the militarism of the Second Reich. John Maynard Keynes, the brilliant English economist, remarked much later, "The German Empire has been built more truly on coal and iron than on blood and iron."[15]

Coal was discovered in the United States by the early explorers and settlers as they spread inland from the first footholds on the eastern seaboard. Coal may have been used occasionally as a local fuel starting in the late 17th century. The first attempt at a coal-mining operation occurred in 1701 near Richmond, Virginia. A commercially successful coal-mining industry dates from about 1745. In the 18th century, wood and charcoal were the principal fuels in the United States. The millionth ton of coal was not produced until 1840, 95 years after commercial mining began. Coal mining increased rapidly thereafter, keeping pace with the rapid development of industry and the spread of railroad transportation. A mine in Pocahontas, Virginia, produced 44 million tons of coal during its 85-year working life. Coal was primarily used for domestic heating and energy for small commercial establishments in areas near the mines. The use of coal in iron and steel making became increasingly important in the late 1800s. Today, the principal use of coal in the United States is for electricity production; metallurgical uses are a distant second, and domestic use is almost nonexistent.

Coal in the Recent Past

The relatively peaceful and prosperous years from the end of the U.S. Civil War to the outbreak of World War I witnessed an enormous growth in coal production. The total production of coal throughout the world increased more than fivefold, from about 180 million tons in 1865 to 930 million tons in 1905. As the 19th century ended, the United States surpassed Great Britain as the world's leading coal producer.

This tremendous increase in the production and use of coal was not sustained. The world consumption of coal was actually the same in 1970 as in 1912.[16] During these years, coal consumption was almost constant, rising somewhat during the war years and falling during the Great Depression and the recessions of the 1950s. Coal production started another upward trend during the 1970s as a result of increasing industrialization in previously underdeveloped countries and as a result of increased interest in coal caused by the oil price shocks.

At the end of World War II, coal was still the predominant energy source in the United States. In 1945, coal supplied 60% of U.S. energy needs. However, a rapid growth in the use of oil and natural gas occurred in the 1950s. These two fuels do have some attractive advantages in comparison with coal. Being fluids, they are easier to handle than the bulky solid coal. They can be supplied to a stove, furnace, or boiler merely by opening a valve—no one has to shovel oil. When oil and natural gas are burned, they leave no ash residue behind to be handled and discarded. They can be shipped by pipeline or easily loaded into and unloaded from trucks, railroad cars, or tankers. The conveniences of oil and natural gas are especially attractive for households. From the late 1940s to the late 1950s, the market for coal used in homes in the United States plummeted by two-thirds, from 99 million tons in 1947 to only 33 million tons in 1958.

Because the 1950s were a time of industrial expansion, coal was able to hold its own in terms of total production. The market lost by homeowners switching away from coal and by railroads converting to diesel locomotives was recompensed by increased use of coal in the utility industry. Total coal use stayed roughly constant, but the ways in which it was being used were changing dramatically. Because the total use of all forms of energy was increasing greatly, the percentage (as opposed to absolute amount) of the total energy consumption supplied by coal dropped steadily. Coal's share of total energy consumption in the United States dropped from 60% in 1945 to about 33% in the late 1950s, to 25% in 1965, and to a low point of about 20% in the late 1970s. The Soviet Union surpassed the United States as the world's leading coal-producing nation in the 1950s.[17] Total consumption of energy in the United States more than doubled from 1950 to the first oil price shock in 1973. Although coal use remained constant, oil use tripled and natural gas use quadrupled.

In the 1970s, the Organization of Petroleum Exporting Countries (OPEC) imposed an oil embargo. Drastic increases in oil prices dramatically illustrated to the United States and other countries the economic and political dangers of relying on an energy source for which a sizable fraction must be imported. Coal usage, which had been dormant during the 1960s, suddenly revived and became an increasingly attractive energy option. In Japan, for example, the amount of coal imported increased by 25% in only 3 years.

The rosy future predicted for coal in the late 1970s and early 1980s has faded somewhat because of recent sharp drops in the price of crude oil and the near disintegration of OPEC as a serious world political force. Even so, growth in coal use in the near future is expected for several reasons:

1. Supplies of oil and natural gas sooner or later will become increasingly scarce.
2. New processes for more efficient coal burning and conversion to synthetic fuels or chemicals will offer new options for using coal.
3. The concern of relying on energy imported from politically unstable regions of the world remains.
4. The increasing development of the Third World has opened new markets for coal and has changed the focus of world coal use.

In regard to the fourth point, a significant shift has occurred in the use of coal throughout the world since 1950. In 1950, North America and western Europe collectively accounted for about 65% of world coal production. Since then, the focus has shifted to eastern Europe, the Soviet Union, China, and the Pacific Basin. That 65% of coal production is now accounted for by China, the Soviet Union, East Germany, Poland, and Czechoslovakia. Only about 30% is produced in the United States and western Europe. In 1984, Australia became the world's leading coal-exporting nation.[18]

Today, more than 1 billion tons of coal are mined and burned each year. Coal supplies about 30% of the world's energy needs (down from 60% in 1950).[19] The major uses of coal are for electric power generation and in the iron and steel industry. In the United States the use of coal in home furnaces now accounts for less than 5% of coal use, but in many parts of the world coal remains an important source of home heat. Process industries, such as the cement, paper, and textile industries, also use coal as a heat source.

No one can predict the future with certainty. Still, because of the inevitable eventual decline in oil and gas supplies, continued research into improved ways of using coal, and continued growth in the total energy demand throughout the world, energy from coal will be an increasingly important component of the world economy for many years to come.

Describing and Classifying Coal

A brief examination of coal samples from around the world shows that the word *coal* actually describes materials with a wide range of appearance and properties. Coal ranges from a soft, moist, brownish material to a very hard, glossy, black solid. To be sure of what we really mean when talking about coal, we must have standard ways of describing and classifying the soft, moist, brown coal, and the hard, glossy, black coal as well as all the varieties in between. Commercial trade in coal depends on the buyer and seller agreeing on exactly what is being traded. The efficient use of coal requires that engineers know something about the specific material they are burning. The traditional way of characterizing coal to meet these needs relies on three kinds of analyses.

Proximate Analysis

The first approach to examining coal is based on the behavior of coal when it is heated. Heating is one of the most easily available ways of bringing about changes in substances. All processes that make use of coal involve heating it. For these reasons, an examination of the effects of heat on coal is a likely starting point for characterizing the various kinds of coal.

If coal is heated gently at 105 degrees Celsius, it loses weight. The material that is removed at this temperature is water. This test is the basis for determining the amount of water in coal. Usually, water is referred to as *moisture* in coal analysis.

After the moisture has been removed, the temperature can be increased. Because coal burns if heated in air, this test is done in an inert atmosphere. Heating coal at several hundred degrees Celsius causes another weight loss. Some of the material that escapes in this experiment is a mixture of gases including carbon dioxide and methane. The remainder condenses as an oily organic liquid and a tar. Ordinarily, the gases, oil, and tar are not analyzed further to determine the amounts of pure compounds present in each. Instead, the total amount of gas, oil, and tar evolution is lumped together and called *volatile matter.*

The solid material remaining after the volatile matter has been removed is a black char resembling charcoal. If air is admitted to the apparatus, the black char burns and leaves behind an inorganic solid called *ash.* The amount of coal material in the char is called *fixed carbon*[20] and is calculated by subtracting the percentages of moisture, volatile matter, and ash from 100.

The set of procedures for determining moisture, volatile matter, and ash and for calculating fixed carbon is known as the *proximate analysis.* Use of this term is misleading because proximate is often thought to be short for approximate. In fact, nothing is approximate about a proximate analysis. The procedures for performing a proximate analysis are very carefully defined and rigorously followed to ensure repeatability from one laboratory to another. The specific details are prescribed by standards-setting organizations in various countries. In the United States, the proximate and other analyses are governed by standards established by the American Society for Testing and Materials (ASTM). The term *proximate* refers to the practice of lumping together a group of constituents and reporting those constituents as one generic item, for example, volatile matter.

The results of the proximate analysis confirm the observation of great variability among coals. The moisture content ranges from nearly zero to almost 70%. Similarly, the ash content varies considerably, from less than 1% to about 30%. Volatile matter ranges from 5% to almost 60%, and fixed carbon ranges from 20% to 85%.[21]

The proximate analysis is relatively easy and quick to perform. It is useful in making empirical predictions of the behavior of coal, particularly during combustion. The only fuel constituents of the coal are the volatile matter and fixed carbon. Moisture and ash represent unreactive components and consequently lessen the amount of heat to be obtained from a given weight of coal. Ash also represents a waste material. It must be removed from the combustor and discarded. The relative proportions of volatile matter and fixed carbon also help to predict how the coal will burn. Volatile materials vaporize and burn as a gas and give a long, smoky flame. Fixed carbon burns as a solid and gives a short, relatively smokeless, hot flame.

Ultimate Analysis

Although the proximate analysis is helpful in practical applications, it doesn't provide information about the actual composition of coal. The elemental composition is determined by a second set of analytical procedures, which collectively are known as the *ultimate analysis*.

The elements that are normally determined in the ultimate analysis are carbon, hydrogen, sulfur, and nitrogen. Carbon and hydrogen are measured by burning the coal and collecting the resulting carbon dioxide and water in chemicals that will absorb them. By weighing the absorbents before and after the analysis, the amounts of carbon dioxide and water can be determined. The carbon in carbon dioxide and hydrogen in water are known with great accuracy. The weights of carbon dioxide and water can then be used to calculate the weights of carbon and hydrogen in coal. For example, 1 gram of carbon dioxide contains 0.273 grams of carbon. Each gram of carbon dioxide made in the experiment must have come from 0.273 grams of carbon in the coal sample. The weight of carbon in the sample and the weight of the sample itself are then used to calculate the percentage of carbon in the sample.

In one of the common methods used to determine sulfur, the sulfur can be converted to sodium sulfate, which in turn can be converted to barium sulfate. Barium sulfate is highly insoluble in water and therefore can be collected, dried, and weighed. Because the amount of sulfur in barium sulfate is known, the weight of barium sulfate can be used to calculate the amount of sulfur in the coal, just as the weights of carbon dioxide and water were used to determine carbon and hydrogen.

Nitrogen is usually determined by converting it to ammonia and absorbing the ammonia in a known amount of acid. The amount of acid remaining after the ammonia has been absorbed is subtracted from the amount used initially. The difference represents the acid neutralized by the ammonia and thus tells the amount of ammonia made in the experiment. Then, the amount of nitrogen present in the coal sample can be calculated.

Relatively simple analytical methods to determine oxygen in coal are not available. The most commonly used method today involves bombarding the coal with neutrons to convert the oxygen to a very unstable isotope[22] of nitrogen (nitrogen-16). The nitrogen-16 decays rapidly by emitting beta (β) and gamma (γ) rays, which can be detected and counted. A comparison of the results from a coal sample with those from a pure compound in which the oxygen content is well-known provides a measure of the oxygen in the coal. Very few laboratories have access to the appropriate neutron source, that is, a nuclear reactor. Many others do not want the expense, time delay, or bother involved in sending samples to a reactor facility. Consequently, the amount of oxygen is determined by adding together the percentages of the other four elements and subtracting this amount from 100. The disadvantage of this procedure is that all of the experimental errors in the carbon, hydrogen, nitrogen, and sulfur measurements are propagated into the calculated value for oxygen.

We have seen that the results of the proximate analysis show considerable variability among coals. The same is true of the results of the ultimate analysis, especially for carbon and oxygen. The carbon content of coal runs from about 70% to 95%. The oxygen content drops as carbon content increases. Oxygen content ranges from more than 25% to almost zero. Hydrogen and nitrogen show little change; the hydrogen content is about 5.5% in most coals. Carbon content can be up to 90%, and nitrogen content is typically about 1%.

Carbon is the element that provides most of the heat energy when coal is burned, but for many people, sulfur is the element of greatest concern, or notoriety, in coal. Burning coal for heat or electric power generation converts the sulfur to sulfur dioxide and sulfur trioxide. These gases can cause air pollution unless they are removed from the combustion products or unless the sulfur is removed from the coal before it is burned. Sulfur occurs in coal in several forms. Some sulfur is part of the molecular architecture of the coal itself. Sulfide minerals, such as pyrite, sometimes deposit in cracks or pores in the coal. The ultimate analysis measures the total sulfur content of the coal without distinguishing among the forms in which it occurs. Total sulfur varies widely, from less than 1% to more than 8%. The portion of total sulfur actually bound to the coal, called organic sulfur, usually stays in a much narrower range, that is, less than about 1.5%. The pyritic sulfur causes most of the variability in total sulfur content.[23]

The ultimate analysis belies its name; it is not really the "ultimate" analysis for determining all the elements in coal. Ash is composed mainly of 10 elements: aluminum, calcium, iron, magnesium, phosphorus, potassium, silicon, sodium, sulfur, and titanium—all combined with oxygen. Analysis of coal down to the smallest concentrations chemists can measure reveals that coal actually contains at least trace amounts of every known element except the noble gases and the highly unstable elements made in nuclear reactors.

Heating Value

A third coal analysis determines the amount of energy that can be obtained by burning coal. This quantity is called the *heating value* or the *calorific value*. In this analysis, a known quantity of coal is burned inside a reaction chamber, which is then immersed in a known quantity of water. The heat released in the combustion of the coal sample raises the temperature of the surrounding water. The amount of heat causing the temperature of a given mass of water to rise 1 degree Fahrenheit is accurately known. The mass and temperature increase in the water and the mass of the coal sample provide the information needed to calculate the heating value. In the United States, the heating value is reported in British thermal units per pound (Btu/lb). A *British thermal unit* is the amount of heat needed to raise the temperature of 1 pound of water 1 degree Fahrenheit.

Like the results of proximate and ultimate analyses, the heating values of coals vary widely, from less than 6000 Btu/lb to more than 14,000 Btu/lb. To compare coal with some other common fuels, diesel oil has a heating value of 19,600 Btu/lb; gasoline has a heating value of 21,400 Btu/lb; and natural gas has a heating value of about 22,300 Btu/lb.

Classification

Because coal exhibits such a wide range of composition and properties, some kind of system is needed to bring a semblance of order from this seeming chaos. At the very least, a set of labels for different kinds of coal could help sort coals into classifications having roughly similar properties. For coal, classification is made on the basis of rank, and the names of the ranks of coal are the labels that provide some information about the coal. For instance, one of the ranks of coal is called *anthracite*. To sellers and buyers of coal and to researchers studying coal, anthracite implies a coal that is very high in carbon content, up to 95%; has a high heating value, perhaps 13,000 Btu/lb; and is low in ash, moisture, and sulfur. Furthermore, many properties vary in a fairly regular way with rank. The oxygen content, for example, drops steadily as rank increases. Thus, the rank of a coal also gives an indication of how the properties of that particular coal relate to other coals having different ranks.

The rank of a coal also indicates the extent of geological maturity. Chapter 2 explains how coal is formed from plants. The final maturation product of coal formation is *graphite,* a form of pure carbon. Rank assigns the position of a particular kind of coal in the progression from the original plant material to carbon. Because the carbon content of coal increases as it matures, rank also gives a qualitative indication of the amount of carbon in the coal.

In the United States the system used for classification of coal into ranks was developed by the ASTM. The system uses the volatile matter and fixed carbon results from the proximate analysis and the heating value as the indicators of rank. The ASTM classification system is given in Table 1.1.

Table 1.1. The ASTM System for Classifying Coals by Rank

Class	*Group*	*Fixed Carbon*[a]	*Volatile Matter*[a]	*Heating Value*[b]
Anthracitic	metaanthracite	>98	<2	
	anthracite	92–98	2–8	
	semianthracite	86–92	8–14	
Bituminous	low-volatile	78–86	14–22	
	medium-volatile	69–78	22–31	
	high-volatile A	<69	>31	>14,000
	high-volatile B			13,000–14,000
	high-volatile C			10,500–13,000
Subbituminous	subbituminous A			10,500–11,500
	subbituminous B			9500–10,500
	subbituminous C			8300–9500
Lignitic	lignite A			6300–8300
	lignite B			<6300

NOTE: This classification system is based on ASTM standard D 388–66, which is published annually by ASTM in their compilation of standards.

[a] The fixed carbon and volatile matter, reported as percentages, are determined on a dry, mineral-matter-free basis. The mineral matter is calculated from the ash content by the Parr formula: mineral matter = 1.08[percent ash + 0.55(percent sulfur)].

[b] The heating value, reported in British thermal units per pound, is expressed on a moist, mineral-matter-free basis. The moisture content is the bed moisture or equilibrium moisture of the coal after equilibration with a nominally 100% relative humidity atmosphere. Some overlap occurs in the heating-value range of subbituminous A and high-volatile C coals. Coals with heating values between 10,500 and 11,500 are classified as high-volatile C bituminous if they display caking properties and as subbituminous A if they do not.

The system derives from an interest in classifying coals for their suitability in commercial use. The heating value indicates the amount of heat to be derived from burning the coal. The volatile matter and fixed carbon help determine the burning characteristics, particularly the nature of the flame.

The first product in the formation of coal is *peat*. Peat is not considered to be coal, but it is an important product in the formation of coal from plant substances. The ranks of coal, in increasing order, are lignite, subbituminous coal, bituminous coal, and anthracite. A distinction is sometimes made between brown coal and lignite, namely, brown coal is less mature. Bituminous coal is subdivided, on the basis of volatile matter, into low-, medium-, and high-volatile groups. (The other ranks are also subdivided into groups, but the group names are not as commonly used as those of bituminous coal.) Lignite and subbituminous coals are known collectively as low-rank coals, and bituminous coals and anthracite are known as high-rank coals.

The Ranks of Coal

Brown Coal

Some distinction can be made between brown coal and lignite. *Brown coal* has a very high moisture content, sometimes more than 60% as it is mined, and it is a younger coal geologically. Deposits of brown coal show distinct layering; the layers are distinguishable by color. When brown coal dries on exposure to air, large pieces tend to disintegrate, a process called *slacking*. When stored in stockpiles or bins, brown coal occasionally catches fire by spontaneous combustion. Brown coal is easy to ignite but has a low heating value, about 3000 Btu/lb as mined. An enormous deposit of brown coal occurs in the state of Victoria, Australia. Brown coal also occurs in central and eastern Europe.

Lignite

Among the coals of the United States, *lignite* is the lowest in rank, containing less than 75% carbon on a moisture- and ash-free basis. The moisture content of lignite is high, but not as high as that of brown coals. The highest moisture content of lignite mined in the United States is 40–42%; this lignite is from the Gascoyne mine in Bowman County, North Dakota. Lignite is relatively soft and ranges from brown to black. Because lignite has not progressed very far in maturation, many lignite deposits contain easily recognizable plant remains up to the size of branches or stumps. The reproductive parts of plants—spores and pollen—are well-preserved in lignite.

Lignite has a high volatile matter content and consequently ignites easily; however, it burns with a smoky flame when used in fireplaces or domestic furnaces. Like brown coal, lignite slacks when it dries on exposure to air and can experience spontaneous combustion in storage or shipment. The heating value is low; it is about one-half that of bituminous coal. For these reasons, lignite was not used extensively until the 1970s. The only use of lignite was in areas of western states very close to mines so that the poorer quality of lignite balanced the high shipping costs of bituminous coal brought in from the East. Lignite use began to increase in the 1970s because of the construction of large power stations directly adjacent to the mines. Electricity is always easier to transport than coal. New power-station construction in the northern Great Plains made that region an exporter of electric power to other parts of the country. The increasing energy demands in the Sunbelt caused a significant expansion of lignite use in Texas.

Two major deposits of lignite occur in the United States. The Fort Union region spreads over North Dakota, Montana, Wyoming, and parts of Saskatchewan and Manitoba. This lignite deposit is the largest in the world. The Gulf Coast lignites stretch from southern Texas, across Louisiana and Arkansas, and into Mississippi and Alabama. Lignite is an important resource outside the United States as well. Some Third World countries, such as Haiti, are turning to lignite as an option for domestic cooking and heating to try to halt the serious deforestation caused by their reliance on firewood.

Subbituminous Coal

Subbituminous coal is intermediate in rank between lignite and bituminous coal. Subbituminous coals have matured to a point at which the woody texture often seen in brown coals or lignites is no longer apparent. Subbituminous coals are black, having none of the brownish color seen in some lignites. Years ago subbituminous coal was sometimes called black lignite. Subbituminous coals have the same tendency toward slacking and spontaneous combustion as lignites and brown coals have. Large deposits of subbituminous coals occur in the United States in the Rocky Mountain states.

Subbituminous coals contain enough volatile matter to be easily ignited, but they are cleaner burning than lignites. For many years, subbituminous coals mainly served a regional market for domestic and small-scale commercial heating needs. No special interest was attached to the fact that subbituminous coals are generally very low in sulfur. Then came the greatly increased concern for environmental quality in the 1970s, a concern that included air pollution caused by the emission of sulfur oxides from coal combustion. The available strategies for reducing emissions were installing systems to remove the sulfur oxides from the combustion products, cleaning the coal to reduce the sulfur content before combustion, or switching to a coal that initially had a low sulfur content. The last option was very attractive because it allowed sulfur emission requirements to be met with little or no additional investment. Subbituminous coals offered a reasonably high heating value—not as good as bituminous coals but better than lignite—and very low sulfur content. The market for these coals exploded. Today, subbituminous coal from the Powder River region in Wyoming is shipped to 20 states for burning in power plants.

Bituminous Coal

Bituminous coal supplies most of the energy that comes from coal. Until the 1970s the low-rank coals provided only about 2% of the national coal consumption. Bituminous coal and anthracite made up the rest; bituminous coal had about a 6:1 margin over anthracite. Bituminous coal heated homes, powered locomotives and ships, smelted iron, generated electricity, fired furnaces and boilers, and made illuminating gas. Even with the changes coming from increased use of subbituminous coal and lignite, bituminous coal is still predominant.

Bituminous coal is black but frequently appears to be banded with alternating layers of glossy black and dull black. It breaks into prismatic blocks. The moisture and volatile matter contents are lower and the heating value is higher than those of subbituminous coal. Bituminous coal shows little tendency to decrepitate on weathering or experience spontaneous combustion.

When some varieties of bituminous coal are heated in the absence of air, they soften. Heating also removes volatile matter as gases, which bubble through the softened mass of coal. The product resolidifies as a shiny, porous, hard, black solid known as *coke*. Coke is the fuel used in blast furnaces to make iron. The discovery of how to make coke led to the widespread availability of cheap iron and was one of the major contributors to the Industrial Revolution. The byproducts of coke making are a rich variety of organic chemicals that were for years the basis of the organic chemical industry.

The bituminous coal with the lowest volatile matter content is sometimes called *semibituminous coal*. It has the highest heating value of any coal. The low volatile matter, and thus the high fixed carbon, means that this coal burns with a very clean flame; thus, it is valuable for domestic use. In the days of coal-fired steamships, semibituminous coal was greatly favored as a naval fuel because it provided the highest amount of heat for the least amount of fuel space.

Bituminous coal occurs in the Appalachian Mountain chain, running from Alabama up into Pennsylvania. A second major deposit lies in the center of the United States, particularly in Illinois. Bituminous coal is widespread throughout the world. The principal countries having bituminous coal include Great Britain, Germany, the Soviet Union, China, and Australia.

Anthracite

Anthracite ranks highest among coals, a position merited for several reasons. Having a very low volatile matter content, anthracite burns with a hot, clean flame with no smoke or soot. This feature makes it an ideal domestic fuel. Compared with bituminous coal, anthracite burns more slowly and gives off heat more uniformly. Anthracite is very low in moisture, about 3%, and is low in sulfur. It is stable in storage and, unlike other coals, can be handled without dust forming. Its heating value comes close to that of the best bituminous coals.

Anthracite does not form coke when heated. Occasionally, it has been blended with bituminous coals in coke manufacturing. However, its hardness, strength, and slow burning make anthracite a good blast furnace fuel. Indeed, anthracite fueled the iron furnaces in the early days of the iron industry in the United States.

Anthracite is jet black and usually has a high luster. It is the hardest and most dense coal. In the days when coal meant only bituminous coal and anthracite, the terms soft coal and hard coal were often used to distinguish the two. Anthracite was occasionally called stone coal. It was also known as black diamond because of its hardness, luster, and commercial value.

The many good qualities of anthracite make it a premium fuel. Unfortunately, its geographic distribution limits its use. In the United States, significant deposits of anthracite occur only in a small region of northeastern Pennsylvania. Therefore, it is not as readily available on the market as are other coals and thus commands a premium price.

Notes

[1] *World Book Encyclopedia*; World Book–Childcraft International: Chicago, 1979; Vol. 4, p 556.

[2] Ibid., p 569.

[3] Ibid., p 581.

[4] Morgan, A. *Getting Acquainted with Chemistry*; D. Appleton–Century: New York, 1942; p 36.

[5] The word *anthrax* is better known to us as the infectious disease of cattle and other animals, which sometimes spreads to humans. Anthrax and anthracite are both derived from Theophrastus' *anthrax*. The connection is that anthrax in humans produces skin sores that have centers as black as coal.

[6] Grainger, L.; Gibson, J. *Coal Utilisation: Technology, Economics and Policy*; Graham & Trotman: London, 1971; p 113.

[7] Gimpel, J. *The Medieval Machine*; Penguin: New York, 1976; p 82.

[8] Several versions of the origin of the term *sea coal* are known. The term may refer to coal that was picked up on the beaches around Newcastle. More likely, it derives from the fact that coal from Newcastle was shipped by sea to London. In any case, sea coal generally is not considered to be high-quality coal because it burns with a smoky flame and does not provide much heat.

[9] *Encyclopaedia Britannica*; Encyclopaedia Britannica: Chicago, 1969; Vol. 8, p 119.

[10] Ibid., p 83.

[11] Ibid., p 84.

[12] In *Henry IV, Part 2,* written about 1600.

[13] *Encyclopaedia Britannica*; op. cit., Vol. 5, p 963.

[14] Manchester, W. *The Arms of Krupp*; Little, Brown & Co.: Boston, 1968; p 86.

[15] Ibid., p 14.

[16] Hottel, H. C.; Howard, J. B. *New Energy Technology*; MIT Press: Cambridge, Mass., 1971; p 5.

[17] *World Book Encyclopedia*; op. cit., p 582.

[18] *1987 Information Please Almanac*; Houghton–Mifflin: Boston, 1986; p 138.

[19] *World Book Encyclopedia*; op. cit., Vol. 6.

[20] *Fixed* is an ancient term in chemical nomenclature meaning resistant to change. Fixed carbon is fixed because it doesn't volatilize. Some gases, such as oxygen and nitrogen, are sometimes referred to as fixed gases because they cannot be made to condense to liquids at temperatures normally attainable in a laboratory.

[21] The data for ash, volatile matter, and fixed carbon are given on a moisture-free basis. The moisture content can be influenced by rain falling on a coal stockpile or a sample standing in a hot, dry laboratory. The other constituents are sometimes reported on a moisture-free basis by multiplying by 100/(100 − moisture). Because the coal substance itself is only in the volatile matter and fixed carbon, these two components can be reported on a moisture- and ash-free basis by multiplying by 100/(100 − moisture − ash). Other analyses can be similarly recalculated to these bases.

[22] Isotopes are chemical elements that have the same atomic number (number of protons) but different atomic mass (due to a different number of neutrons).

[23] A third component contributes to the total sulfur, the so-called sulfate sulfur, which consists of various sulfates. In most coals the sulfate sulfur is a minor component.

Chapter 2

How Coal Was Formed

Coal was formed from plants by the chemical and geological alteration of plant material over tens or hundreds of millions of years. The geochemical process that transformed plant remains into coal is called *coalification*. Scientists from many disciplines, including chemistry, geology, and botany, have contributed to our knowledge of coalification; their collective efforts are part of the general area of research known as *organic geochemistry*.

Much evidence supports the conclusion that plants were the starting point for the coalification process. The first piece of evidence comes from consideration of the chemical composition of coal. The element that predominates in coal is carbon. The carbon content of coal ranges from 60% to 95% by weight. For every 100 atoms in a bituminous coal of 84% carbon by weight, 55 atoms would be carbon, 40 would be hydrogen, and about 4 would be oxygen, as well as 1 or 2 atoms each of nitrogen and sulfur. The only likely source of vast amounts of material rich in carbon and hydrogen and containing some oxygen, sulfur, and nitrogen would be living, or once living, tissue. By itself, this observation does not necessarily exclude animal remains as a precursor of coal.

The examination of coal deposits readily provides evidence in favor of plants. Fossilized plant remains can be found associated with coal of all ranks. Seams of lower rank coals, particularly brown coal and lignite, occasionally are found to contain wholly or partially coalified pieces of plants, ranging in size from sticks and branches to logs. A tree branch found in a lignite seam is shown in Figure 2.1. Furthermore, when coal is examined under a microscope, remains of a cellular structure very similar in appearance to plant tissue can be seen. Microscopic examination also

Figure 2.1. The oval structure pointed out by the pencil tip is the cross section of a tree branch occurring in a lignite seam. (Courtesy of David Kleesattel.)

reveals a diversity of pollen grains and spores. Figure 2.2 shows pollen from a modern pine tree, and Figure 2.3 shows pollen from a lignite sample. Both pollen samples are from the same species of tree.

A few coal seams contain unusual deposits called *coal balls*. These coal balls are petrified, rather than coalified, aggregations of plant material that have been preserved by impregnation with pyrite or calcium carbonate. Coal balls vary considerably in size, but a typical coal ball might be roughly the size and shape of a grapefruit. To the miner, coal balls are a nuisance to be discarded as waste, but to the geochemist, coal balls are valuable objects for study because they provide an almost perfect preservation of the details of the cells of the original plants.

A technique frequently used to probe the chemical composition of coal is the treatment of coal samples with solvents to dissolve, or extract, various constituents of the coal. Identifying the compounds in the solvent extract provides some information about the molecular architecture of the coal. These experiments also help to relate the coal to its parent plant material. Among the substances that can be removed from coal by solvent extraction are ones that may be identical to compounds occurring in plants today or, more likely, have molecular structures closely related to components of modern plants. These kinds of compounds, which have survived coalification virtually intact, are sometimes called *biological markers* because they are compounds that help to identify, or mark, the biological origin of the samples being studied. By understanding the chemical changes that transformed the plant constituents into biological markers, geochemists can

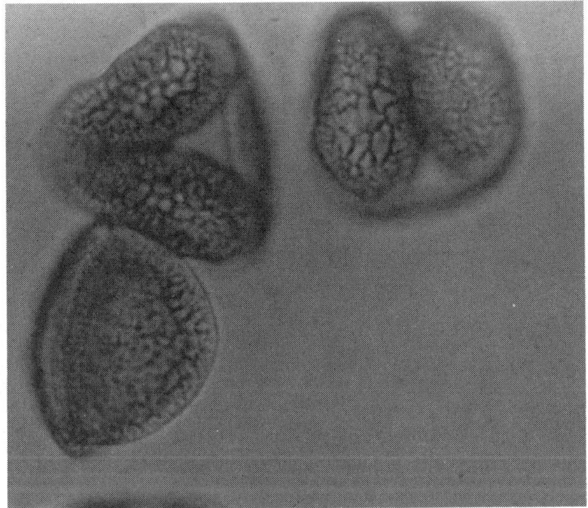

Figure 2.2. Pollen from a modern pine tree. (Courtesy of Edward Steadman.)

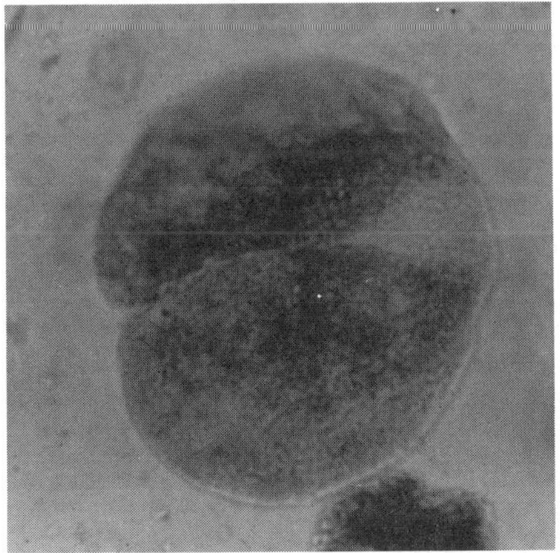

Figure 2.3. Pollen from a lignite sample. This pollen and the pollen in Figure 2.2 are from the same species of tree. This pollen is approximately 60 million years old. (Courtesy of Edward Steadman.)

infer the kind of plants contributing to the coal, some of the details of the chemistry of coalification, and details of the environment in which the coal was deposited. For example, many plants secrete *resins*, which contain a family of compounds called *terpenes*, as well as related diterpenes and triterpenes. Extraction of lignites has yielded triterpenes, which are characteristic of compounds found in flowering plants, and diterpenes, which are indicative of conifers. Relating the biological markers in coal extracts to chemical constituents of plants provides additional evidence for the botanical origin of coal.

Plants That Made Coal

The next step in understanding how coal was formed is to determine the kinds of plants and environments that contributed to what eventually became coal. To do so requires an inquiry into the history of the earth since the development of plant life. Geologists have divided the earth's history into five eras, three of which are further divided into periods. This system of describing the history of the earth is called the *geologic time scale*[1] and is outlined in Table 2.1.

Fossilized plant remains have been found in Precambrian rocks from such places as Michigan and Ontario. Although these materials are of interest and importance to paleobotanists, they do not constitute workable coal deposits. Coal geology really begins in the Devonian period.

Silurian and Devonian Periods

The first plants to grow on land made their appearance toward the end of the Silurian period. These were simple plants—slender, leafless shoots—but were the earliest representatives of the *vascular plants,* that is, plants characterized by tubes or channels that serve to transport nutrients and water throughout the plant structure. Vascular plants were a major evolutionary advance in plant life. Although plants did not appear on land

Table 2.1. The Geologic Time Scale

Era	*Period*	*Approximate Age (millions of years)*
Cenozoic	Quaternary	11
	Tertiary	70
Mesozoic	Cretaceous	135
	Jurassic	180
	Triassic	225
Paleozoic	Permian	270
	Carboniferous	350
	Devonian	400
	Silurian	440
	Ordovician	500
	Cambrian	600
Proterozoic		2500
Archean		4600

until about the end of the Silurian period, extensive forests existed by the middle of the Devonian period. The oldest coal comes from rocks about 360 million years old, that is, from the close of the Devonian period. During the Devonian period, two groups of plants became increasingly important. One was a group whose descendants today are the club mosses; the other was a group including scouring rushes and horsetails. Many of these early plants inhabited subtropical forests and preferred moist places. This evidence suggests that the environments in which coal was deposited were warm and moist.

Carboniferous Period

The plants of the Devonian period underwent a great expansion both in geographic range and in size during the Carboniferous period. The Carboniferous period derives its name from the enormous deposits of coal that formed at that time. In the context of coal geology, the Carboniferous period is perhaps more extensively studied than all of the other periods collectively. Alexander Winchell luridly described the transition from the Devonian period to the Carboniferous period in his book *Walks and Talks in the Geological Field:*[2]

> *While the grizzly monsters of the ancient deep were luxuriating in empire and blood, the premonitions of progress were felt. The world was not made for them, but only an age of the world. Behold, the tide bears out into the sea a floating log. Its exterior is marked by peculiar and significant impressions. They reveal the crest of a dynasty in the vegetal world. It has floated from the shore of a low-lying and silent continent. There is a meager nursery there where nature is training vegetation for thriftier times. In this log is written the doom of the old Placoderms which had stirred the Devonian mud of the Ohio sea-bottom. It is a voice crying in the wilderness of waves, "Prepare the way for new land, new forms, new scenes, and new history." It was the beginning of the Carboniferous age.*

During the Carboniferous period, the world climate was warm and humid, typical of what we would consider tropical or subtropical regions today. The weather was uniform throughout the year: long growing seasons; moderate-to-heavy rainfall; and little likelihood of severe, crippling frosts. The most familiar modern analogue of the Carboniferous climate in which the coals formed might be the climate of the coastal plain region of Florida up to about the Carolinas. These conditions were ideal for a proliferation of plant life. The plants of the Carboniferous period grew over vast tracts of land and assumed enormous proportions compared with their modern descendants. For example, the Carboniferous club mosses were the size of trees, but modern club mosses are small plants. More than 3000 species of plants have been identified as contributing to coal formation from fossilized or coalified tissue, pollen, or spores.

Ferns were especially important members of the Carboniferous forests. Today, when we see ferns growing in a garden or in the wild, we are unlikely to think of them as having any particular economic value. However, the ancient progenitors of these seemingly insignificant plants contributed to the formation of untold billions of tons of coal, which in turn have provided us with heat and power for transportation, electricity, and the steel and chemical industries. About 300 species of ferns flourished during the Carboniferous period; in comparison, only about 50 modern species are known in the United States today. These ancient ferns grew to heights of 40–100 feet and had stems 4 feet in diameter. (To put these figures in perspective, an impressively large elm or oak shade tree averages about 100 feet tall).

Today, the ground pines and club mosses are small plants; they occasionally grow to 3 feet in tropical forests, but often trail no more than a few inches off the ground in temperate regions. The ancestors of the ground pines and club mosses, called scale trees, were among the most prolific inhabitants of the Carboniferous forest; they were often the dominant vegetation. Large specimens reached 125 feet, although 30–75 feet was a more common height. More than 40 species have been identified. The name derives from the scars left on the stem where leaves were attached. One genus had a spiral pattern of scars and a branching structure somewhat like an elm; another had scars longitudinally up the stem and no branching, probably looking something like a 100-foot-tall bottle brush. The stumps of these trees are sometimes found as fossils in the rock beneath the coal seam. Like the ferns, the scale trees reproduced by spores. The spore cases were cone-shaped and sometimes attained a length of 18 inches.

Giant forerunners of today's horsetail and scouring rushes were abundant and were the size of modern trees; they grew in thick stands (dense forests). As is true for the ferns and scale trees, the descendants are few in number, about 25 species, and they are small. In some places, these rushes may grow to 1 foot in height, but their Carboniferous ancestors sometimes attained 1 foot in diameter.

These plants all reproduced by spores. The first seed plants of the class that includes the conifers appeared during the Carboniferous period. By the late Carboniferous period, the class had undergone great expansion so that conifers were important contributors to the plant life of the period. The expansion of the conifers and other seed plants occurred at the expense of the spore-bearers.

The Carboniferous forests were enormous. A single coal deposit dating from the Carboniferous period is the Pittsburgh seam of bituminous coal.

The Pittsburgh seam occurs in southwestern Pennsylvania, northern West Virginia, and southeastern Ohio in an area of 20,000 to 25,000 square miles. An equivalent forest today would cover the entire land area of West Virginia or all of New Hampshire and Vermont as well as some of Massachusetts.

This profusion of plant life flourished in a swampy environment. During the Carboniferous period, swamps covered great areas. The swamps were on flat, low-lying ground along the shores of shallow seas or near the mouths of rivers. The water was fresh or brackish in most cases. Normally, the movement of surface water in the swamp was minimal. The swamps were occasionally subjected to extensive flooding or complete drying. The swampy environment was important for the preservation of plant material, and the flooding or drying had a role in determining the extent of coal deposition.

Permian Period

Most of the major deposits of bituminous coal and anthracite in the Northern Hemisphere accumulated during the Carboniferous period. Coal deposition was reduced significantly in the Permian period, except in China and Manchuria. In the Southern Hemisphere, however, the important deposits of bituminous coal in South Africa and Australia were laid down in the Permian period. The plant life in the Permian period was much the same as in the Carboniferous period: ferns, scale trees, and giant horsetails. Also, early conifers and their relatives appeared during the Permian period. However, the preserved plant tissues show evidence of growth rings. A plant accumulates growth rings when its biological activity is not continuous but occurs only in periods determined by seasonal changes. This evidence suggests that the perpetual summer of the Carboniferous period had ended and that the climate was cooler and experienced seasonal changes.

Cretaceous Period

Coals of the Triassic and Jurassic periods are not widespread. In terms of total coal production and use worldwide, they are of minor importance. The next major period of coal formation was the Cretaceous period. Coals of the Cretaceous period occur in many parts of the world from Spitzbergen to Antarctica. Coal occurring in such seemingly inhospitable places tells us that the climate in those regions was once much milder than it is today. Enormous reserves of low-rank coals of the Cretaceous age occur in the western part of North America. Important deposits also occur in Australia, New Zealand, Germany, and Siberia.

The Cretaceous period was marked by a significant shift in vegetation. During the early part of the Cretaceous period, ferns were still abundant and conifers had undergone extensive development. (Conifers belong to a class of plants called *gymnosperms,* a term that means "naked seeds". This term refers to the exposed seeds in pine cones.) Then, a new kind of plant began to develop. This new class enclosed its seeds in a protective covering, for example, inside an apple. The plants with enclosed seeds are called *angiosperms.* As plant evolution proceeded during the Cretaceous period, the angiosperms became increasingly important, and they are the dominant plant form today.

Tertiary Period

Deposition of coal-forming plant material continued into the Tertiary period, and angiosperms were the major contributors to the coal deposits of this period. Ashes, willows, and maples were among the trees that were flourishing. A reconstruction of the depositional environment of lignite in western North Dakota, based on a study of pollen and spores isolated from the coal, suggests that the environment was an extensive cypress swamp, somewhat like our modern Okefenokee Swamp in Georgia and Florida.

Components of Plants

The coalification process involves the conversion of the chemical components of plant tissues into coal. To understand something about coalification, we need to begin with some knowledge of what these chemical components are. Plant components can be divided into three categories:

1. structural components, that is, chemicals contributing to the rigidity and form of the plant;
2. stored foods, which are used as energy sources by the plant; and
3. specialty materials, which provide a variety of services to the plant.

Structural Components

The most common structural material is *cellulose.* Cellulose is made by the plant by the polymerization of glucose (grape sugar or corn sugar). Cellulose is found in cell walls, providing the basis of plant rigidity, and can compose up to 60% of plant material. In woody tissue, cellulose makes up the bulk of the tissue on a dry-weight basis. In the chemical industry, it is the starting material for making such products as cellophane and rayon. Cellulose is gradually destroyed during coalification. It is found in peat and to lesser extents in brown coal and lignite. Cellulose is generally absent from higher rank coals.

Lignin is the second most abundant structural material. Its presence in the cell wall adds rigidity to the structure. In some plants, old cell walls made originally of cellulose gradually become impregnated with lignin. Lignin is also produced by polymerization, but its structure is much more complex than that of cellulose. The carbon atom arrangement in the polymer is derived from *n*-propylbenzene. Lignin is more resistant than cellulose to decomposition and therefore tends to persist longer during coalification. Brown coals and lignites may contain about 10% lignin.

When two plant cells come in contact, they produce a wall between them. This middle wall is made of *calcium pectate.* Most of us occasionally bite a piece of fruit that has become overripe and pulpy; this condition is caused by the disintegration of the middle wall and, consequently, the loss of some of the structural rigidity that the middle wall had provided. The mushy consistency results from the conversion of calcium pectate to *pectin.* We take advantage of this phenomenon in cooking by adding pectin to help break down fruit structure when making jams and jellies.

Hemicellulose is also a polymer made from sugar structures. It differs from cellulose in that the starting material for hemicellulose is *xylose* (wood sugar) and in that the hemicellulose molecules are about one-thousandth the size of cellulose molecules. Hemicellulose is usually found in the cell walls of seed and stem tissues. Unlike cellulose and lignin, hemicellulose is sometimes used as a reserve food supply by the plant.

Stored Foods

Carbohydrates are the most important of the food materials. Simple sugars, such as *glucose,* are the first products in the synthesis of food by the plant. The simple sugars are subsequently used to make more complex structures, which are stored by the plant for eventual use as food. The most important of the complex carbohydrates is *starch.* The carbohydrates are relatively unstable chemically, in comparison with cellulose and lignin, and consequently decompose very early in the coalification process.

Fats and *oils* are other important energy sources for the plant. The distinction between the two is simply a physical one: fats are solids at normal temperature, whereas oils are liquids. Fats are very efficient energy reserves because a molecule of fat contains proportionately less oxygen for a given amount of carbon than the carbohydrates. The conversion of stored food to energy is the process of *oxidation;* having less oxygen, fats are able to liberate proportionately more energy than carbohydrates.

Proteins are a ubiquitous component of the living material filling the cells. Proteins are metabolized as food when the supplies of carbohydrates and fats have been exhausted. All plant components are compounds of carbon, hydrogen, and oxygen. These are the only elements in the structural components and the carbohydrates. Proteins, however, contain nitrogen and sulfur as well. The nitrogen and some of the organically bound sulfur in coal probably come from the plant proteins.

Specialty Materials

Resins are one of the specialty materials in plants. Resins are found particularly in conifers; one way to become intimately acquainted with resins is to take down a Christmas tree. Conifer wood and bark tissue has specialized ducts where resins accumulate. In some plants, the resins accumulate at points where the plant has been injured, so they serve a protective function. Resins are quite stable during the early stages of coalification. Eventually, the resins tend to polymerize to hard solids such as rosin or amber.

Cutin is secreted in the outer wall of leaf cells. It acts as a seal, protecting the interior regions of the leaf from the atmosphere. Cutin is another material that tends to survive early coalification.

Waxes are compounds made from fatty acids and alcohols or sterols. They are synthesized in the outermost cells of the plant—the *epidermis*—and help seal the moisture in the plant. Like cutin and resins, waxes endure through coalification. In fact, some lignites contain enough wax that it is profitable to mine the lignite simply to extract the wax. The lignite-derived wax is used as a constituent of shoe polish, furniture polish, and waterproof paints.

Sporopollenin is a polymer found in the protective outer layers of spores and pollen. The chemical structure of sporopollenin is not known, although a consensus among coal scientists is that the molecular structures proposed in the past are almost certainly wrong. Sporopollenin is very resistant to change during coalification. Although it is not a major constituent of most coals, it may help us learn more about the coalification process. When the molecular structure of sporopollenin is determined, the structure of sporopollenin from modern plants can be compared with that found in pollen or spore samples isolated from coal. The structural differences between the two could provide clues about the chemical reaction conditions, temperatures, and pressures that were experienced during coalification.

Photosynthesis and the Carbon Cycle

The ultimate source of the carbon in the various plant constituents is carbon dioxide in the atmosphere. Plants capture carbon dioxide with the help of sunlight in the process known as *photosynthesis.* During photosynthesis, plants form the simple sugars. Then, with these sugars available as nutrients, plants can form the other protoplasm components, cell walls, and specialty materials. When plants use their nutrients for energy, some of the carbon is returned to the atmosphere as carbon dioxide. When the plant dies, its tissues decompose, and thus carbon dioxide is again returned to the atmosphere. If the plant is eaten by an animal, or if the plant-eating animal is eaten by another animal, carbon dioxide is returned to the atmosphere by respiration of the animals while they are alive or by decay of their bodies after their deaths.

The process by which atmospheric carbon dioxide is fixed by plants, carbon compounds are used by plants or animals, and carbon dioxide is returned to the atmosphere by respiration or decay is called the *carbon cycle.* This cycle is shown in Figure 2.4. In principle, the carbon cycle gradually moves all the organic carbon in the world from the atmosphere, through living beings, and back to the atmosphere. The reason that we have enormous deposits of coal in the earth is that the carbon cycle is not 100%

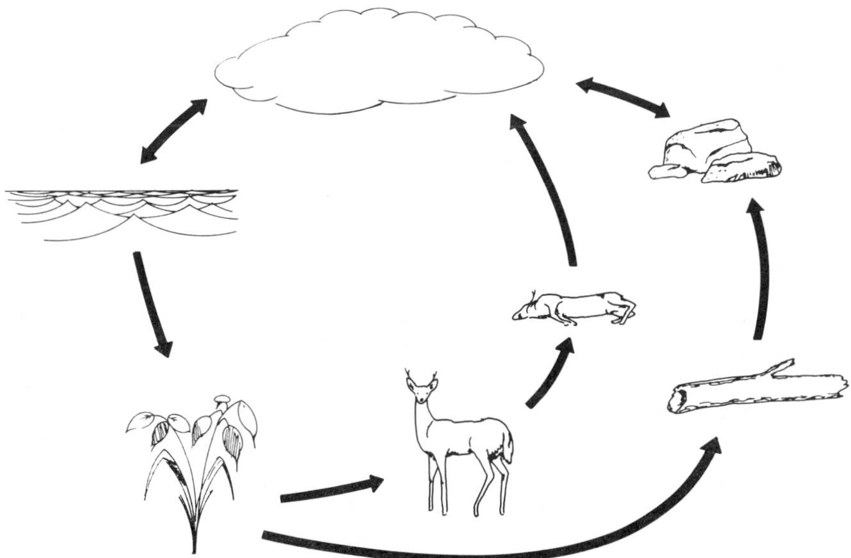

Figure 2.4. In the global carbon cycle, carbon dioxide in the atmosphere is absorbed by growing plants. The plants may die and decompose, or they may be eaten by animals that eventually die and decompose. The decomposition processes return the carbon to the atmosphere as carbon dioxide. The carbon dioxide in the atmosphere is in equilibrium with dissolved carbonates in the ocean and carbonates in rocks. The formation of coal results when the decomposition of plants is interrupted. (Drawing by Gwyn Schobert.)

efficient. Starting in the Devonian period, some small fraction of the land plants, maybe 5%, did not fully decay. Instead of being converted all the way back to carbon dioxide, this small bit of material experienced a totally different fate. It turned to coal.

From Plants to Peat

The plant communities that ultimately formed coal deposits grew in a swampy environment. Direct evidence for this conclusion is provided from preserved root structures that resemble those of plants growing in water. Indirect evidence is provided by the continuous gradation of composition and properties that occurs among peat and the various ranks of coal; also, all modern deposits of peat are in swampy or boggy environments. The watery environment was the factor responsible for introducing the slight inefficiency in the carbon cycle that permitted the formation of coal.

Plant Decay

When a plant dies in a relatively dry environment, such as in a forest, decay starts immediately. The plant tissue is gradually converted to carbon dioxide and water as well as small amounts of ammonia and sulfur compounds. Eventually nothing is left except the inorganic components of the original plant tissues. The principal chemical agent in the decay process is atmospheric oxygen. When a plant falls into water, however, the water shields the dead plant from oxygen in the air. The principal agents causing decay are *anaerobic bacteria,* that is, bacteria able to live and function in the absence of oxygen. The anaerobic bacteria decompose the plant material by removing hydrogen and oxygen through the gaseous compounds methane and carbon dioxide, respectively, and as water. Even though some carbon is lost in the methane and carbon dioxide, much more hydrogen and oxygen are lost than carbon. Thus, the net effect of this decomposition process is to build up a residue that is rich in carbon. (For example, cellulose contains 165 atoms of hydrogen and 83 atoms of oxygen for every 100 atoms of carbon, but lignite contains only about 88 hydrogen atoms and 20 oxygen atoms per 100 carbon atoms.)

Effect of Water Environment

The decomposition process eventually stops because the bacteria kill themselves. Some of the decomposition products produced as intermediates prior to the formation of carbon dioxide, methane, and water are organic acids and phenols. In the stagnant water of the swamp, these acids increase in concentration to a point at which the bacteria can no longer survive. In rushing water, such as in a rapidly moving river, the organic acids would be diluted and swept away. Thus, the bacteria would be allowed to continue their attack on the plant material. The swamp environment, therefore, serves a double purpose: (1) The water itself provides a shield that protects against attack by atmospheric oxygen, and (2) the relative quietness of the water allows the accumulation of substances that halt the anaerobic decay by killing the bacteria.

A freshwater environment is the most favorable and most productive for eventual coal formation. The coal-producing plants may grow along meandering streams or flood plains, in bogs, or even in aging lakes having high enough concentrations of plant nutrients to support profuse plant growth. Brackish environments are not quite as favorable but still have supported plant communities that led to extensive coal deposits. River deltas or coastal swamps are examples, as are bays or lagoons that have some protection from vigorous water movement. Marine environments are the least favorable. Nevertheless, some large coal seams originated in marine conditions, including the bituminous coals in the central part of the United States.

Even in a swampy environment, coalification does not necessarily proceed unhindered. If the swamp is flooded, the flood waters may dilute the toxic species accumulated in the water, just as a river would. Also, the flood waters may wash away the plant debris and thereby remove the components of the coal. On the other hand, if the swamp dries out when the water level is lowered, the plant debris is exposed to the air, and the resulting action of oxygen leads to complete decomposition. If either the flooding or the drying out persist for long, coalification is effectively halted. Instead, some other inorganic sediment may be deposited on top of the newly forming coal. At a later time, the swamp may be reestablished and thus lead to the formation of a new deposit of coal. This cyclic process of swamp formation and destruction by flooding or drying eventually results in a series of coal seams being deposited and separated by other, inorganic sediments.

Biochemical Phase of Coalification

The bacterial attack on the plant material is the first step in the formation of coal. This process is sometimes referred to as the *biochemical phase* of coalification. The bacteria first attack the carbohydrates, starch, and pectin. Hemicellulose and then cellulose are the next to go. Lignin is more resistant than cellulose. Resins and waxes are even more resistant to bacterial attack, and sporopollenin is probably the most resistant of all. Leaves and other soft tissues slowly disintegrate into a variety of residual materials. Almost total disintegration produces a black *muck*[3]. Some of the plant debris retains its structure because of the comparative resistance of lignin to decomposition.

If the water in the swamp is brackish or saline, enough sulfate can be in the water to support bacterial life that converts sulfate ions to hydrogen sulfide or sulfur. These two products can be further incorporated in the plant residues as organic sulfur or as growths of pyrite. The incorporation of sulfur in either form has profound environmental consequences because when the coal is burned to release its energy, the sulfur is converted to sulfur dioxide. Without appropriate controls, the sulfur dioxide would escape to the atmosphere and eventually form sulfurous and sulfuric acids. Our concerns today for the effects of acid rain and the costs of incorporating pollution-control equipment in power plants stem from seemingly insignificant bacteria quietly attacking sulfate ions in swamp water 300 million years ago.

Peat Formation

As more plant debris accumulates, the material on the bottom of the pile becomes compressed. The compressed plant tissue and residual organic debris, complete with sulfur and whatever inorganic species were in the plants or were washed in, is *peat.* The formation of peat represents the end of the biochemical phase of coalification. Up to this point, the process could even be referred to as "peatification".

Peat provides visual evidence for the transformation from plant material to coal. The appearance of samples of peat varies considerably; it can be a relatively light-colored, yellowish brown substance in which tangled pieces of plants can be clearly seen to be compacted, or it can be a dark brown or black mass that resembles a wet brown coal or lignite. As masses of dead plants accumulate, layers of vegetation in various stages of decomposition pile up. The weight of the upper layers compacts the material on the bottom and thereby squeezes out some of the swamp water to consolidate the mass of material into a more dense state. Bacterial attack on the cellulose in the plant cell walls helps to disorganize or disintegrate the large pieces of plant tissue. As peatification proceeds, the carbon content increases from about 50% in wood to about 60% in peat. One of the products of the bacterial reactions is methane; sometimes the methane can actually be observed bubbling up through the water. Methane is sometimes called *marsh gas* because of its formation from decaying plant material in relatively stagnant swamps or marshes.

The formation of peat is a relatively quick process on the time scale of geological events, where time is counted in thousands or even millions of years. It is, however, a slow process by human perception. It takes about a hundred years to form a layer of peat 2–3 inches thick. Some modern swamps have peat layers in excess of 30 feet; this condition suggests that growth, death, and decay have been proceeding relatively undisturbed for more than 15,000 years. The ancient environments that led to coal formation were tropical and subtropical regions of luxurious vegetation having stagnant or slow-moving water, as illustrated in Figure 2.5. In the United States at least three places satisfy these criteria and may be modern examples of the peat swamps of the Carboniferous or Cretaceous periods. These places are the Okefenokee Swamp in Georgia and Florida, the Everglades in Florida, and the Great Dismal Swamp in Virginia and North Carolina. The Great Dismal Swamp is a cypress swamp similar to those that flourished in the Cretaceous and Tertiary periods; it contains about 0.75 billion tons of peat.

In most classification schemes, peat is not considered a form of coal. Nevertheless, research on the geochemistry of peat is an important component of coal research because the transformation of plant material to coal proceeds through the formation of peat and then through the subsequent conversion of peat to coal. Peat is sometimes used as a fuel, particularly in countries such as Ireland[4], which have few other indigenous sources of energy. When first harvested, peat contains about 90% moisture, so it must be dried before use. Even when dried, it is not as good a fuel as coal. Peat has been classified as a renewable resource; however, it accumulates so slowly that even if no irreparable harm were done to the peat-forming environment by the harvesting operation, the formation of a new peat bed would take centuries.

No distinct division occurs between the biochemical phase of coalification and the subsequent events, which are referred to as the *geochemical phase.* The formation of peat represents the transition between the two phases. The plants died, the bacteria attacked, and a layer of peat was established. All this took a few thousand years. Now biology stops and geology begins. Millions of years will pass as the processes of the earth slowly but inexorably transform the consolidated peat into coal.

From Peat to Anthracite

A change in the environment can destroy a swamp and halt the formation and accumulation of peat. For example, the sea level could rise, and a higher sea level would allow the sea to advance into the swamp and drown the plants. When this change occurs, layers of new sediments begin to build on top of the peat. The weight of the sediments lying on top of the peat bed creates a pressure that slowly compacts the peat. As more sediments accumulate, the peat is not only compressed by their weight, but it is also buried deeper. This process exposes the peat to elevated temperatures. The combined effect of temperature and pressure is to transform the peat to brown coal and lignite.

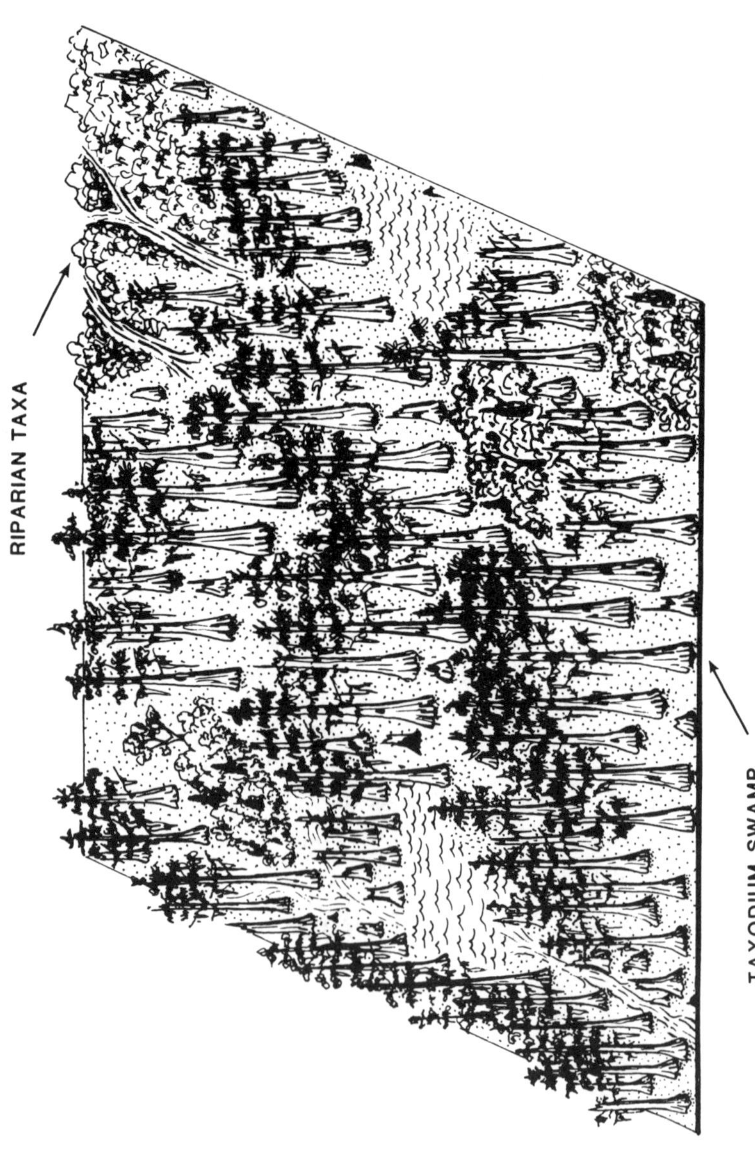

Figure 2.5. Conceptual drawing of the environment of lignite deposition in North Dakota about 60 million years ago. The taxodium swamp is an environment like the modern Okefenokee Swamp; riparian taxa are plants living along rivers. (Drawing courtesy of Edward Steadman.)

Brown Coal and Lignite Formation

Both brown coal and lignite contain recognizable plant fragments. Pieces of leaves, seeds, wood fragments, branches, and even fruits are commonly found in these coals. Partially coalified tree stumps or logs are also occasionally seen. Coalification decreases the moisture content and increases the carbon content. Harvested peat may contain 90% moisture, whereas the moisture contents of brown coal and lignite as mined are 60–70% and 30–40%, respectively. Peat is about 60% carbon (on a dry, ash-free basis); lignite is about 70% carbon.

Bituminous Coal Formation

Bituminous coal forms as temperature and pressure continue to exert their effects over thousands of years. Compared with lignite, bituminous coal has significantly more carbon, up to about 88%, and has lost much of the oxygen originally in the molecular structures of the lignin and other plant components. Peat contains 30–35% oxygen, and lignite retains about 20–25%. In bituminous coals the oxygen content may be only about 7%. Some of the differences between lignite and bituminous coal may reflect differences in the original plants that formed the coal. Lignite was deposited in the Cretaceous and Tertiary periods, when the conifers were abundant and the angiosperms were beginning to become a significant factor in the plant communities. The bituminous coals, however, were deposited during the Carboniferous period, when the giant spore trees dominated. Although the different plant ingredients provided by the swamps of the Tertiary and Carboniferous periods may be a factor in some of the differences between lignites and bituminous coals, a counterargument is provided by the remarkable coal deposit found in the Mahakam Delta in Indonesia. Peat continues to accumulate in the Mahakam Delta. Under the peat, to a depth of about 13,000 feet, is a sequence of coal seams of gradually increasing rank. The lowest seam is a coal of about 85% carbon. The modern peat and virtually all of the coal underneath have apparently derived in large part from a single kind of small palm tree.

Although lower rank coals are often of Cretaceous or Tertiary age and higher rank coals are often of Carboniferous age, rank does not necessarily correlate with age. The temperature and pressure to which the coalifying material was exposed are the key factors; temperature is the primary agent responsible for change during the geochemical phase of coalification. The principal source of heat is the thermal gradients in the earth. The longer and deeper a coal has been buried, the more it has been exposed to the earth's heat and the more extensively it has been coalified. Extremes of temperature, such as may be caused by an intrusion of magma near a coal deposit, can result in a relatively young coal being transformed to high rank.

Deposits of brown coal and lignite found near Moscow were formed during the Carboniferous period. These coals were never buried deeply and thus never experienced the natural driving forces of temperature and pressure that would have transformed them to bituminous coals. On the other hand, small deposits of anthracite in the Rocky Mountains are very young geologically but have experienced very high temperatures.

Anthracite Formation

Anthracite is the highest rank of coal and geologically the oldest. The formation of anthracite involved enormous pressures that were exerted on the coal seams and surrounding rock layers during the formation of the Appalachian Mountains. This event occurred during the Permian period and is sometimes known as the Appalachian Revolution. It resulted from the collision of Africa with North America. The crust of the earth is composed of about 6 large "plates" and about 14 small ones, which very slowly, over the course of geological time, wander over the globe. At the end of the Carboniferous period, the continents had drifted close enough together to form one supercontinent, *Pangaea,* shown in Figure 2.6. As the continental plates moved around to weld Pangaea together, the large "bulge" of western Africa—Sierra Leone, Senegal, and Guinea—collided with the East Coast of North America. If any humans had inhabited Pangaea, the changes probably would have been imperceptible over the course of a lifetime. On a geological time scale, however, the continents slammed together with such ferocity that the rocks themselves folded and buckled. The remarkable patterns of folding can still be seen in rocks exposed by road cuts along any of the major highways crossing the Anthracite region of eastern Pennsylvania. The

Figure 2.6. Most of the earth's land mass was once concentrated in a single supercontinent called Pangaea. The subsequent fragmentation of Pangaea produced the continents as we know them today; the precursors of North America, South America, and Africa should be evident in this sketch.

heat and pressure generated by the folding produced the anthracite. Anthracite is about 95% carbon. Virtually all traces of the original plant life are gone.

Effects of Pressure and Temperature

The geochemical phase of coalification is, in essence, the application of pressure and temperature over millions of years. The combination of pressure and time progressively eliminates hydrogen and oxygen through water, methane, and carbon dioxide. These transformations cause the carbon content of the coal to increase steadily as coalification proceeds; carbon content is 50% in wood, 60% in peat, 70% in lignite, 75–85% in bituminous coal, and about 95% in anthracite. The oxygen content drops sharply during the transition from lignite to bituminous coal, although at that stage the hydrogen is preserved and may even show a slight increase. Finally, a significant reduction in hydrogen occurs as anthracite is formed.

Increasing temperature, or temperature applied over increasingly long times, resulted in increased coalification. Temperature and pressure increases can result from the piling of successive layers of new sediments, including new layers of peat or coal, onto a coal seam. If several seams of coal are found in a vertical sequence, the one at the bottom is the oldest and the most extensively coalified. Pressure increases can also come from the folding of rock during the raising of mountains. Such an event sharply escalates the effects of pressure and temperature and carries coalification to the extreme of anthracite.

A description of coalification may sometimes sound like an experiment in synthetic organic chemistry performed on a grand scale. If chemical A is mixed with chemical B in solvent C and heated at a certain temperature for a certain time period, the product will be chemical D. Similarly, if peat is buried at a particular depth and therefore experiences a particular pressure for millions of years, the product will be a bituminous coal of a certain quality. Unfortunately, this simplification neglects the fact that coals are like snowflakes or human fingerprints: No two are exactly the same in every aspect.

Effects of Chemical Composition of Plant Material

Most likely, variations in the chemical composition of the original plant material carry through coalification and contribute to the variability in coal composition. The isolation and identification of biological markers have been especially useful in developing an understanding of the relationships between coals and their parent plant material. In particular, studies of a family of compounds known as *triterpenes*[5], which are very widely distributed in nature, have provided evidence that the kinds of plants and peat from which coals are formed, as well as geochemical processes, greatly determine the chemical natures of the coals. The kinds of sediments surrounding the coals are also important. Some materials, such as clays, do not resist pressure well and deform relatively easily when pressure begins to build. Thus, a coal deposit in contact with clays may reflect the results of pressure more than other, originally similar coals lying between more resistant materials. Different plants growing in different environments subjected to greater or lesser degrees of pressure and temperature may result in two coals that have, on gross examination, the same apparent rank and carbon content. Yet, a detailed study of these coals will certainly reveal significant differences in properties.

Characterization of Coal Deposits

When a particular coal deposit is sampled and characterized, the nature of the sample reflects the following factors:

- character of the original plants;
- prevailing conditions when the plants died and began to form peat;
- extent of bacterial attack during the biochemical phase; and
- pressure, temperature, and length of time of the geochemical phase.

The conditions prevailing during the geochemical phase are probably the most important factor. Also, variations from the prevailing conditions must be considered. For example, was the coal subjected to moderate temperature and pressure early in the geochemical phase? Did the coal experience severe conditions for a relatively short time? To explain the composition and characteristics of a coal sample is a detective assignment of extraordinary complexity. Progress requires patiently assembling clues from widely different areas such as the biochemistry and anatomy of plants; the extraction and identification of complex organic molecules (biological markers); the geology of sedimentation; and the history of the sinking, rising, and folding of rock layers. The detective–scientist who accepts the assignment will be extremely challenged, but the rewards of intellectual stimulation and satisfaction are indeed worthy of the pursuit.

Minerals Mixed with Coal

When we look at coal with a magnifying lens or microscope, or sometimes with the unaided eye, specks of minerals can be seen. When we burn coal, an inorganic residue is left behind, formed by the action of the fire on minerals or other inorganic constituents that must have been in the coal before it was burned. How these inorganic substances accumulated while the organic plant tissue was being converted to coal is also a part of the coalification story.

Sources

Inorganic constituents become incorporated during coalification in several ways. The most straightforward sources of inorganic substances are the plants themselves. As part of their life processes, plants accumulate some inorganic matter. This phenomenon is seldom noticed in living plants, and we may think of it only when we see the accumulation of wood ash in a fireplace or campfire. Some plants, however, build up appreciable concentrations of inorganic material. An example is the scouring rush, a rather tiny descendant of the gigantic treelike rushes of the Carboniferous forests. The outermost cells of the rush (in the epidermis) are strengthened by deposits of silica. The silica not only strengthens the cells of the living plant but also provides a gritty texture. This texture made the plants very useful to Indians and pioneers as scrubbing or scouring brushes, hence the name.

Once the plants died and fell into the water, they were surrounded by a medium that is an excellent transport agent for dissolved inorganic ions and for bits and pieces of mineral grains. Peat, as well as brown coal and lignite, is able to capture dissolved inorganic matter from the water by a process known as *ion exchange.* The accumulation of inorganic matter starts at the beginning of the biochemical phase of coalification. The composition of the inorganic matter acquired in this way will be determined by the following factors:

- composition of the rocks from which the waterborne inorganic materials were leached or eroded,
- patterns of water flow to and through the peat, and
- chemical ability of the peat to remove ions from solution.

Much of the inorganic chemistry of a coal is fixed by the time it completes the peat stage.

Another source of inorganic constituents of coal is waterborne grains of minerals. Water slowly moving through the coal swamp may carry small mineral particles washed away from rocks surrounding the swamp. These mineral particles can settle and become mixed with the coalifying plant material. Eventually, these particles are incorporated into the coal as discrete grains of minerals called *detrital minerals.* Clay and quartz are examples of detrital minerals.

As coalification continues, the coal becomes increasingly compressed by the overlying sediments and thus becomes less porous or permeable to water flow. Brown coals and lignites are still subject to penetration by water (some seams of these coals are excellent aquifers), and their structures still contain numerous acidic sites for ion exchange with water. Therefore, the amount of exchangeable ions, such as sodium and calcium, in these coals is probably continually changing as the composition and flow patterns of water moving through the seam change. A consequence of this ability to exchange ions is that brown coals and lignites may exhibit great variability in their inorganic composition, whether measured vertically or laterally through the seam. When coalification has proceeded to the bituminous coal stage, the acidic sites in the coal are no longer present (the loss of these groups contributes greatly to the sharp drop in oxygen content that occurs between lignite and bituminous coal). Thus, the capture or exchange of ions, which began early in the biochemical phase, ends with the transition to bituminous coal.

Even after the coal has been buried and compacted and has lost most of its original water content, some water can still percolate through cracks or joints in the seam. Minerals can sometimes precipitate from the water and be deposited in the cracks. A mineral that can get into coal in this way is *pyrite,* the source of much of the sulfur in bituminous coals.

Sulfur in Coal

When coal is analyzed carefully for both major and trace constituents, almost every element will be found. About the only exceptions are the highly radioactive elements that are made artificially and some of the noble gases. Of the roughly 80 elements found in coal, the one besides carbon having the most significant effect on coal use is sulfur. When coal is burned, the sulfur oxides formed cause air pollution if allowed to escape into the environment. The sulfur in coal has caused the spending of millions of dollars on equipment and processes to capture the sulfur oxides or to reduce the sulfur content of coal before it is burned. Sulfur occurs in coal in three forms:

1. organic sulfur, which is a part of the molecular composition of the coal itself;
2. pyritic sulfur, which occurs in the mineral pyrite and some related minerals; and
3. sulfate sulfur, which is mostly iron sulfates.

In most coals, sulfate sulfur is a small fraction of the total sulfur content. In low-rank coals, organic sulfur may contribute half or more of the total sulfur, but in most bituminous coals, the majority of the sulfur is pyritic sulfur.

Some of the organic sulfur is a remnant of plant tissue. Proteins in plants contain 1.0–1.5% sulfur, and some of this amount survives coalification. Much of the sulfur, however, is a result of bacterial activity. One of the products of bacterial attack on sulfur-containing compounds is hydrogen sulfide. This compound is the most noticeable product of bacterial attack (rotting) on sulfur compounds in eggs. The hydrogen sulfide reacts with iron compounds in the swamp water to precipitate iron sulfides, which are gradually converted to pyrite. One of the principal sources of the sulfur is the sulfate ion. In fresh water the concentration of sulfate is very low, but in saline water the sulfate is second in amount only to chloride among the anions. These high sulfate concentrations have profoundly affected the sulfur contents of coals in the United States.

The coals found in the Appalachians were deposited under conditions that changed occasionally from fresh water, when coal deposition occurred around a large inland sea, to brackish or saline water, when the inland sea became an embayment of what is now the Gulf of Mexico. Because the sulfate contents of the water varied accordingly, the sulfur content of the coals in Appalachia also varies, running from about 0.5% to more than 5%. The bituminous coals in the interior of the United States were deposited from saline water at a time when a large sea effectively split North America in half. Because of the saline environment, these coals are high in sulfur; some Iowa coals contain as much as 8% sulfur.[6] In and near the Rocky Mountains the coals were deposited along rivers or in river deltas or were protected from the sea by ridges along the coastline. These coals are therefore very low in sulfur. The increased coal use that followed the oil price shocks of the early 1970s, in addition to the heightened concern for environmental quality that occurred about the same time, suddenly transformed these low-sulfur western coals into a national, rather than regional, resource.

Other minerals besides pyrite accumulate in coal after the biochemical stage. These minerals precipitate in cracks or fissures that form after the coal has been hardened and compacted. Quartz, for example, may be liberated by adjacent rocks and recrystallize in the coal. Kaolinite and calcite may also deposit in the coal in this way. In some cases the mineral precipitates even fill cavities that are the coalified remnants of plant tissue. Whether or not these minerals occur in coal depends on the reactions occurring in nearby sediments and also depends on the water flow. Thus, again, many factors contribute to the considerable variability of inorganic constituents in coals.

Coal as Solar Energy

Coal formed from a sequence of biochemical and geochemical transformations of plants that grew millions of years ago. These plants grew by capturing the energy in sunlight, which drives the process known as photosynthesis. During photosynthesis, carbon dioxide and water are converted to carbohydrates, which are used to make lignin, cellulose, and other materials necessary to sustain plant life. When we burn coal to obtain energy for ourselves, we are essentially reversing this process. The coalified plant components are converted back to carbon dioxide and water, and energy is released. In a sense, the energy that is released to us in the combustion process is the solar energy that was locked away slowly and patiently by countless generations of plants flourishing in the coal swamps tens or hundreds of millions of years ago.

Notes

[1] The *geologic time scale* is much more detailed than the condensed version presented here. The subject is discussed more fully in standard textbooks on geology or earth history.

[2] Winchell, A. *Walks and Talks in the Geological Field*; Chautauqua Press: New York, 1886.

[3] In everyday usage the word *muck* has a negative connotation, but in the specialized world of scientists it is a perfectly respectable word. It refers to a material containing black, extensively decomposed plant material.

[4] Ireland is sometimes nostalgically called the "auld sod". Sod, which we usually think of as the green material unrolled to make a new lawn, derives from a Middle Dutch word "sode", which refers to the wet bogs where turfs of peat were cut for fuel. In a sense, then, Ireland is the "old peat".

[5] An example of a triterpene occurring in modern plants is oleanolic acid, which is found in such diverse sources as cloves and birch bark.

[6] This exceptionally high sulfur content makes these coals totally unacceptable for commercial use today without extensive cleaning or sulfur capture. At one time, mining these deposits for sulfur rather than for coal was considered.

Chapter 3

Coal in the World

A complete geological survey of the world does not exist; thus, nobody really knows how much coal the world contains. Obscure parts of the world may contain coal deposits that are completely unknown. The *reserves* of coal, which are the coal deposits that have been carefully measured by detailed exploration, amount to 1.5 trillion tons. Coal reserves, however, are only a fraction of the estimated coal resources. The ultimate resources of coal are all of the coal that is known to exist (the reserves) plus the coal that is inferred to exist from geological observations. The starting point for estimating coal resources is data obtained from exploratory drilling, from natural or man-made outcrops (such as road cuts), and from surveys for other resources such as minerals, water, oil, or gas. Usually the locations of the preliminary discoveries, for example, outcrops or water wells, are widely scattered. Thus, assumptions about the thickness and continuity of the coal deposits must be made. As data accumulate, the validity of the assumptions can be checked, and the estimated resources can be revised upward or downward. The best guess of coal resources in the world is about 12 trillion tons.[1] For our immediate energy needs, the most important number is the amount of coal that is economically recoverable by our current mining technology. The estimate of economically recoverable coal includes assumptions of the depth to which coal can be readily mined, the minimum thickness of the seams that can be mined, and the fraction of the coal that can be extracted. The depths, thicknesses, and percentages used in the assumptions vary from country to country. Therefore, on a worldwide basis, the total amount of economically recoverable coal is affected by variations in the way the number is calculated. With this caveat, we can estimate that the amount of known, economically recoverable coal in the world is 750 billion tons.[1]

If all 750 billion tons of economically recoverable coal could be mined at once and loaded into standard 100-ton capacity railroad cars, the resulting train would be 71 million miles long and would stretch from Venus to Mars. If this stupendous coal train could go 50 miles per hour, a traveler at a grade crossing waiting for the train to go by would have to wait 570,000 years for this train to pass.

Coal is not uniformly apportioned among the world's nations. The Soviet Union, the United States, and China together have 87% of the world's coal. Canada, West Germany, Australia, and Great Britain collectively have another 10%. Most of the world's coal is found in the Northern Hemisphere. This distribution reflects the facts that (1) most of the land mass in the world is in the Northern Hemisphere and (2) most of the world's developed nations, which are the most likely to have conducted resource exploration programs, are also in the Northern Hemisphere.

The importance of coal in a nation's economy does not directly reflect its share of the world's coal. For most of the 19th century, Great Britain led the world in coal production, yet it has less than 1.5% of the coal resources. Australia, possessing not much more coal than Great Britain, is now the leading coal-exporting nation in the world. Poland, with little more than 0.5% of the world's resources of coal, nevertheless relies on its coal-export trade to obtain badly needed hard currency. South Africa has less coal than Poland, yet it produces from coal virtually all of the liquid fuels (and many byproduct chemicals) needed by more than 28 million people.

Soviet Union

Estimates of the coal resources of the Soviet Union vary greatly, but scientists generally believe that the Soviet Union has more coal than any other nation. The estimated and inferred coal reserves may amount to as much as 7 trillion tons, which would be more than half the coal in the world. About 90% of this coal is in Siberia. In the Soviet Union, coal is the energy source in greatest supply. The Soviet Union has become the world's leading coal-producing nation.

Donets Basin

Although most of the coal reserves are in Siberia, the principal deposits being actively worked are closer to areas of high population density and industry. The Donets Basin, lying north of the Sea of Azov in the Ukraine, has been nicknamed the coal scuttle of the Soviet Union. This region produces one-fourth of the total Soviet output of coal; production is about equally divided between bituminous coal and anthracite. The bituminous coal is used for manufacturing coke for the steel industry, obtaining byproducts from coking, and domestic and industrial heating. Coal seams in the Donets Basin are relatively thin and difficult to mine, but the high quality of the coal repays the effort.

The Donets Basin played a role in world affairs. On December 18, 1940, Adolf Hitler signed a document that would change the course of history. The document had the innocuous title of Directive Number 21, but it is best known as Operation Barbarossa—the German plan to invade Russia. Directive Number 21 said, in part, that "pursuit of the enemy will have the following aims: in the south the early capture of the Donets Basin, important for war industry..."

In 1941, the Donets Basin was producing 60% of the coal and 75% of the coke used in the Soviet Union. However, events would prove that the capture of this region would not mortally wound the Soviet industrial system. When the tide of the war turned in 1943, Hitler would not accede to recommendations from field commanders in the Ukraine to fall back to more defensible positions. He felt that even the fraction of Donets coal production that would be retaken by the Red Army would be crucial to the German war effort. The fierce battles that raged for possession of Donets coal cost thousands of lives on both sides.

Kuznetsk Region

The Kuznetsk region ranks second in importance among Soviet coal fields. The region lies between the Trans-Siberian Railway and the Ob River Basin. Kuznetsk Basin reserves are 4–5 times higher than those of the Donets Basin. The seams are thicker, averaging almost 7 feet, and the coal is of good quality. The iron and steel industry uses much of the Kuznetsk coal.

Ural Mountains

Coal occurs in scattered deposits throughout the Ural Mountains. Both bituminous coals and lignites are mined in the Urals. The mining operations are generally easy, but the bituminous coal is not as high-quality as the Donets or Kuznetsk coals.

Kazakhstan Region

Kazakhstan—the region bounded by the Urals, the central Asian republics, and Siberia—produces large quantities of coal for coke production. In northern Europe, coal is mined in the Pechora region. These mines were opened during World War II to replace some of the coal production lost when the Germans captured the Donets Basin. Lignites and brown coal are mined in the Moscow Basin. These coals are important both for economic and scientific reasons. Most coals deposited during the Carboniferous period have metamorphosed to bituminous or anthracite rank, but the coals in the Moscow Basin, which are of Carboniferous age, have not. Evidently, over millions of years, these deposits have not been exposed to significant thermal gradients or tectonic stresses in the earth.

Siberia

Huge reserves of coal exist in Siberia. So far, production from the Siberian coal fields has been relatively small because of the small population in Siberia and the lack of major industrial centers. An important Siberian mining region is the Kazakhstan coal field, which produces excellent coal. This coal is shipped west to the industrial regions in the Urals that need coal for steel production and other metallurgical operations, such as copper smelting. As the Soviet government continues with plans to develop Siberia, the coal fields will become increasingly important.

United States

The coal reserves of the United States are second among nations. The United States led the world in coal production throughout most of the 20th century until the Soviet Union surpassed it in the 1950s.[2] About one-fourth of the world's coal reserves are in the United States. The U.S. Geological Survey has divided the reserves geographically into seven provinces. The provinces are divided into regions and fields. The coal reserves are also roughly divided by rank; most of the bituminous coal and anthracite is in the East, and most of the lignite and subbituminous coal is in the West.

Eastern Province

The Eastern province includes large deposits of bituminous coal and some anthracite. It stretches from the northeastern portion of Pennsylvania southwestward to northern Alabama. The Eastern province is about 900 miles long in a northeast to southwest direction, and almost 200 miles wide at its broadest. The coal deposits of particular interest include the Anthracite field in northeastern Pennsylvania and the Appalachian region, which begins in Pennsylvania and runs through West Virginia into Kentucky, including parts of southeastern Ohio and western Virginia.

The Anthracite field is only about 500 square miles in extent, yet the quality of coal is so high and the deposits are so rich that more than 20% of all the coal ever mined in the United States has come from this small part of Pennsylvania. One vein of anthracite, the Mammoth seam, averages about 40 feet in thickness and has been worked throughout the region, from

Carbondale to Shamokin. Anthracite is found only in places where the sequence of rock and coal layers has been subjected to intense temperature and some increased pressure from rock folding. Because of the folding, the anthracite seams are generally tilted at large angles, sometimes as much as 80° from the horizontal. The tilting of the beds complicates the mining process, making it particularly difficult to rely on the large mining machines used in bituminous coal mines that have more nearly horizontal seams. Except for a brief resurgence during World War II, the Anthracite region of Pennsylvania has suffered chronic depression since the 1930s. Today, anthracite contributes only a small fraction of the total coal production in the United States.

The Appalachian field is one of the greatest deposits of bituminous coal in the world. Much of the field is composed of a vertical stacking of about 50 separate coal seams (117 seams have been identified in West Virginia). About 10 of these seams are being mined. The greatest is the Pittsburgh seam, which covers an elliptical area of roughly 20,000 square miles. It lies in western Pennsylvania, southeastern Ohio, and northern West Virginia. About 5000 square miles of the Pittsburgh seam are being actively mined; in this area the seam ranges from 6 to 14 feet in thickness. An estimate of the tremendous amount of coal in this deposit can be gauged from the rule of thumb that each foot of seam thickness yields 1 million tons of coal per square mile. So far, almost 5 billion tons of coal have been taken from the Pittsburgh seam.

When the coal was being deposited during the Carboniferous period, this region was at times an extension of the Gulf of Mexico, and at other times it was cut off from the Gulf and existed as an inland sea. This change in environment while the coal was being deposited, from saline water of an arm of the Gulf to the almost fresh water of an inland sea, has caused significant variability in the coal of the Appalachian field. For example, the sulfur content ranges from a low of 0.5% to about 6%. The deposition of plant debris over such a wide geographical region resulted in the plant material not being buried uniformly throughout the region. Because coalification depends on the depth of burial, the coal seams in the Appalachian field also show some variability in properties due to this factor.

Today, the Eastern province is enormous, and most likely the original coal deposit was much larger. The anthracite that remains is the portion that was preserved by being folded downward into layers of rock as the mountains were formed 250 million years ago. The rest has vanished forever as a result of 200 million years of erosion.

Further to the southwest, the bituminous coal region of Kentucky became notorious for the fierce and violent labor struggles during attempts to unionize the mines in the 1930s. Harlan County was nicknamed "bloody Harlan" and has figured in literature and folk songs. Many of the intellectuals of the time became sympathetic to the union cause, and their articles and essays attracted national attention to the region. Despite this dismal history and the continual spasmodic outbreaks of violence associated with strikes, Kentucky is now the leading coal-producing state.

Interior Province

Three regions make up the Interior province: the Northern region, in Michigan; the Eastern region, in Illinois, southwestern Indiana, and western Kentucky; and the Western region, comprising parts of Iowa, Missouri, Nebraska, Kansas, Arkansas, and Oklahoma. The Eastern region is the most important commercially.

The Eastern region occupies an area of roughly 50,000 square miles. Illinois has more coal resources than any other state; coal lies under about two-thirds of its area. In terms of sales, coal is the most important of Indiana's minerals. Production of Interior province coals in western Kentucky adds to the production from the Eastern province to give Kentucky the national lead in coal mining.

Interior province coals were mostly deposited in shallow seas. They have not been exposed to the same degree of geological forces as the coals of the Eastern province. Consequently, they have not matured to the same extent and tend to be of lower rank than Eastern province coals. However, the seams are relatively horizontal and close to the surface, so mining tends to be easier. Deposition in sea water, where the sulfate ion is an abundant anion, has resulted in the Interior province coals being generally high in sulfur. Some Iowa coals contain 6–8% sulfur and were once viewed as a potential source of sulfur rather than energy.

Gulf Province

The Gulf province stretches from Alabama and Louisiana into Texas. It runs in a wide belt southwestward through Texas to the Rio Grande. The Gulf province coal is lignite. For many years this resource was used on a minuscule scale because of readily available oil and natural gas in this region. Recently, the continued growth of population and industry in the Sunbelt has resulted in a sharp increase in the use of lignite as the fuel in electric power plants. Texas has become the nation's leading producer of lignite, and virtually all of the commercial development of Gulf province lignites occurs in Texas.[3] Activities occurring in most of the other states of the Gulf province will eventually lead to the commercial use of lignite in those states, also.

Northern Great Plains Province

The Gulf province is one of two major lignite deposits in the United States; the other is the Northern Great Plains province. The Northern Great Plains province lies mostly in western North Dakota and eastern Montana, as well as in parts of northeastern Wyoming and northwestern South Dakota. The coal deposits of the Northern Great Plains province continue across the border into Canada. The Fort Union region is the largest deposit of lignite in the world.

The story of lignite development in the Northern Great Plains province is very much like that of the Gulf province. Until the early 1970s, lignite was used on a very small scale. The Northern Great Plains is an area of low population density and comparatively little industrialization. Energy needs were easily met by small power plants. In the 1970s, the oil embargo and resulting rapid escalation of oil prices, as well as the increased concern for environmental quality, greatly affected the use of energy resources. Electricity made from lignite provided an alternative to using oil or gas in power plants. Lignites as a rule are low-sulfur coals. Shipping lignites long distances is not practical because of their high moisture contents and low heating values. Electricity, however, is easy to "ship"; consequently, a number of large power plants were built in the 1970s to take advantage of the vast deposits of low-sulfur lignite. Several of the plants were built along the Missouri River north of Bismarck, North Dakota, to capitalize on the easy availability of water from the river. As a result, the Northern Great Plains province has become an exporter of electric power to the rest of the country.

The Northern Great Plains province is also home to the nation's first commercial plant making substitute natural gas from coal. Built near Beulah, North Dakota, the Great Plains Gasification Associates plant converts lignite from the Freedom Mine to methane for addition to natural gas pipelines. The plant came on-line in July 1985, and from a technical standpoint it is entirely successful. The operation of the plant has been beset by economic difficulties and political and legal wrangling between the original owners and the U.S. Department of Energy. Its long-term future is uncertain.

The lignites of the Fort Union region were deposited in swampy environments that were much like our modern Okefenokee Swamp. The Gulf province lignites were deposited in saline or brackish environments. Thus, as a rule, the Gulf province lignites tend to be higher in sulfur and ash than the Northern Great Plains lignites. However, because high moisture contents and low heating values make long-distance transportation of lignites impractical, the two varieties do not compete head-to-head in the marketplace.

Rocky Mountain Province

The Rocky Mountain province, as the name implies, contains the coal fields in the mountains running roughly north and south from New Mexico through Colorado, Utah, and Wyoming to Montana. Virtually every rank of coal can be found somewhere in the Rocky Mountain province. Part of the great variability in rank comes from the coal deposits being exposed to varying degrees of thermal stress, for example, from localized magma intrusions, and to varying degrees of tectonic stresses from mountain building. Rocky Mountain province coals are higher in rank than Northern Great Plains coals of the same geological age because of increased thermal and tectonic stresses.

The most important Rocky Mountain coals are in the Powder River region. These are subbituminous coals of low sulfur content occurring in some of the thickest seams in the country. Because subbituminous coals have higher heating values and lower moisture contents than lignites, they are much more suitable for long-distance transportation. Power plants converting to the use of low-sulfur coals in the 1970s caused great expansion of coal production in the Powder River region. Coal from this region is now used in about 20 states. Shipments of coal from the Powder River region to other parts of the United States are shown in Figure 3.1.

Pacific Coast Province

Little coal-mining activity is found in the Pacific Coast province. Subbituminous coal is mined in Washington for power generation. This coal has a high

Figure 3.1. In the late 1970s, the combined effects of the oil embargo and the concern for reducing environmental pollution led to a rapid increase in demand for the low-sulfur coal of the Powder River region. Coal from Wyoming was shipped to utilities as far away as Ohio, Kentucky, and Arkansas.

ash content, frequently containing more than 10% ash and sometimes more than 20%. Some lignite is mined in California for the main purpose of extracting waxes from it. These waxes, which are remnants of the plants that decomposed to produce the coal, are used in shoe and furniture polishes. Pacific Coast coals have not been used extensively, partly because of competition from better quality Canadian coal to the north and California oil and gas to the south.

Alaskan Province

The extensive coal deposits of the Alaskan province contain virtually all ranks of coal. So far, Alaskan coal has been commercially developed only on a small scale because of a low population density, the lack of extensive industrialization, and the availability of indigenous oil and gas and imported Canadian coal. Also, Alaskans have been concerned about environmental problems that could result from widespread strip mining in its fragile ecosystem.

China

China ranks third among nations in terms of total coal resources, and within China, coal is the most important mineral resource. Much of the Chinese coal is bituminous, but extensive deposits of lignite and some anthracite also occur. The province of Shansi, which lies west of Beijing, contains about half the total resources. Other provinces in which coal mining is important are Heilungkiang, located in far north China in the large bow of the Amur River boundary with the Soviet Union; Kirin, across the Yalu River from North Korea; Hopeh, the province in which Beijing is located; and Liaoning, on the northern edge of the Yellow Sea. All of these regions are in northern China, which accounts for about 70% of China's coal.

Most Chinese mines produce coals in the range of millions of tons per year, not unlike large mines in any other coal-producing nation. However, for years the internal transportation system in China lagged behind those of other large nations. To try to have ample supplies of coal available throughout the country and to meet a national goal of local self-sufficiency, an attempt was made to have communes run their own small-scale mines. This effort occurred during and shortly after the notorious Cultural Revolution of the late 1960s. In some areas these coal mines are responsible for as much as one-third of the coal production and have indeed contributed to coal self-sufficiency.

Coal in China presents an interesting blend of the old and the new. China is the last major nation to rely heavily on steam locomotives for a large share of its railroad traffic. However, as contacts with the West improve, China is increasingly interested in obtaining modern technology for coal gasification and importing the latest techniques for coal analysis. The total size and quality of Chinese coal resources are still not well-known in the West, and possibly not in China either. However, as China continues to modernize, coal will certainly be a vital part of its economy.

Canada

The Soviet Union, the United States, and China represent the world's "big three" in terms of coal resources. A second group of nations—Canada, West Germany, Australia, and Great Britain—have smaller shares. Nevertheless, even a small percentage of the world's total resources represents enough coal to have a significant effect on both domestic energy use and exports.

In terms of all possible energy resources—coal, petroleum, natural gas, uranium as fuel for nuclear reactors, and water power for hydroelectric generation—Canada is very likely the most endowed nation. All ranks of coal are found in Canada. The coal in eastern Canada, that is, Nova Scotia and New Brunswick, is bituminous coal used for both power generation and coke making. The eastern Canadian coal fields must be considered with the Eastern and Interior provinces in the United States as major deposits of bituminous coal. Bituminous coal is also the principal type of coal in British Columbia, which is on the other seaboard. British Columbian coals are also used for power generation and coke making. In addition, British Columbia has lignite deposits and small amounts of anthracite. Inland, Alberta has both bituminous coals and lignites. Saskatchewan has large deposits of lignite, which are an extension of the lignites in the Northern Great Plains province of the United States.

Although Canada possesses almost 5% of the world's estimated coal resources, Canadian coal has never figured prominently in world commerce, as has the coal of Great Britain. Also, Canadian coal has not played a role in a major world event, as has the coal of the Donets Basin in the Soviet Union. However, for a brief time in the fall of 1958, attention focused on the town of Springhill, Nova Scotia. On October 23, a cave-in at the Cumberland mine trapped 174 men. Within a few days, 81 men were rescued; another 12 men were found alive after 7 days. Incredibly, 7 more men were rescued on the ninth day; however, the rest died in the accident.[4]

West Germany

Ruhr Region

West Germany contains rich deposits of coal in the Ruhr region and in the Saar Basin. The region where the Ruhr River flows into the Rhine and the

Rhine then turns northwestward into the Netherlands may contain more high-quality bituminous coal than the rest of Europe. The availability of bituminous coal for conversion to fuel for the iron and steel industry, as well as the availability of other raw materials, resulted in the Ruhr becoming one of the most heavily industrialized areas of the world. The Ruhr was a chief means of support for the German economy well before the beginning of the 20th century. The revitalization of the Ruhr following the devastation of World War II was an important component of the "Wirtschaftswunder"—the economic miracle that transformed a defeated and devastated West Germany into one of the world's foremost industrial nations.

For about 80 years, from the late 1860s to 1945, German ambitions and designs shaped events on the continent of Europe. Recall Maynard Keyne's observation that "the German empire has been built more truly on coal and iron than on blood and iron."[5] The passing of that era was accelerated by U.S. and British bombings in 1944 that caused extensive disruption of Ruhr coal transportation to other regions of Germany. Because much of the German rail system relied on steam locomotives, the loss of access to coal from the Ruhr had a crippling effect on rail transportation. Coal was also being converted to liquid fuels to substitute for petroleum. Even some margarine and other edible fats were being derived from coal through a sequence of processing steps. The military collapse of Nazi Germany was hastened by loss of the coal because coal was a principal resource for steel production, transportation, and oil substitutes.

Saar Basin

The Saar Basin lies in the southwestern corner of West Germany, along the borders of France and Luxembourg. This region is important mostly for its coal, and it is the other principal area of bituminous coal mining in West Germany. The Saar contains about one-fourth of the West German coal reserves, which is in itself more coal reserves than many other European nations, most notably France.

The rich deposit of coal lying on the French border was for years a sore point in European relations. During the Peace Conference of 1919, Georges Clemenceau, the premier of France, demanded that the Saar be relinquished to France as part of the German reparations. France had never had the extent or quality of coal reserves of its antagonist to the east, and, to make matters worse, many of the existing French mines had been flooded by the Germans while they occupied French territory. Clemenceau's demand to be given a portion of German territory that had been under French control for at most 25 years led to bitter confrontations with Woodrow Wilson. The issue of the Saar and its coal led to savage attacks on Wilson in the French press and probably undermined his influence on the peace-making process. Eventually, the leaders compromised by giving nominal sovereignty over the

Saar to the League of Nations; however, actual administration of the mining was placed under the de facto control of France. Allied reparations also required large quantities of coal from the Ruhr. The effective loss of the Saar and the delivery of large tonnages of Ruhr coal as reparation payments made it very difficult for German industry to rebuild after the war. This situation, combined with harsh French treatment of the inhabitants of the Saar, contributed to feelings of bitterness and revenge in Germany and sowed the seeds for the next war. On his deathbed, Clemenceau turned to face Germany and warned, "They will come again."[6] He was right.

Coal Adjustment Law

In the 1960s, hard times hit the West German coal industry. Throughout the world, a general shift from coal to oil and natural gas occurred. Some of the German mines that had been worked for decades were now so far underground (sometimes exceeding 3000 feet in the Ruhr) that their operation was becoming increasingly uneconomical. In the 1960s, U.S. coal exports to West Germany were seriously affecting the West German economy. U.S. coal was actually cheaper to the Germans than the Ruhr coal. In 1968, the West German government enacted the Coal Adjustment Law, which provided for closing uneconomical mines. The major coal companies joined into a single holding company, Ruhrkohle AG[7].

Brown Coal

West Germany also has an extensive brown coal resource. The principal deposit lies west of Cologne. The brown coal lies much closer to the surface than the bituminous coal does in the Saar and Ruhr. Thus, brown coal is obtained by strip mining. Combustion in power plants uses about 85% of the brown coal. Another 8–10% is made into briquettes for domestic or small commercial heating needs.

East Germany

East Germany has large brown coal deposits. The principal mining activities occur near Dresden and near Leipzig. East Germany is the world's largest producer of brown coal. As in West Germany, most of the coal is used in power stations for electricity generation. Some is made into briquettes for home heating. Like brown coal in the rest of the world, the East German brown coal is close to the surface and is readily extracted by strip mining. It also has the high moisture content and low heating value characteristic of brown coals. Thus, it is not as desirable for some applications as bituminous coal. In particular, brown coal cannot be converted to coke for metallurgical applications.

East Germany has very little high-quality bituminous coal. Some bituminous coal is mined around Dresden and around Karl-Marx-Stadt (formerly Chemnitz) in the Erzgebirge (the "Ore Mountains") near the border with Czechoslovakia. These deposits do not contain enough bituminous coal to meet the needs of East German industry. Production of bituminous coal is declining, and most of the bituminous coal and coke used in East Germany are actually imported from Czechoslovakia, Poland, and the Soviet Union.

Australia

Australia is endowed with major deposits of both bituminous coal and brown coal. The major coal fields in Australia lie along the eastern and southeastern edges of the continent: Bituminous coal is found in Queensland and New South Wales, and brown coal is found in Victoria. Smaller deposits occur in Western Australia at Collie and in South Australia at Leigh Creek. Coal accounts for more than one-fourth of the total value of mineral production in Australia and is the major source of electric power.

Bituminous Coal

The bituminous coal of Queensland and New South Wales includes sizable quantities of coking coal. As a result of increased industrialization of Pacific nations since World War II, especially within the past 10–15 years, Australian bituminous coal has become an attractive commodity. Japan, in particular, has long imported coking coal from Australia. Because of the oil embargoes and price increases of the 1970s, Japan realized that access to imported oil was dependent on unstable politics and economics. Consequently, the Japanese began importing Australian coal for fuel as well as for coke. In the past few years the energy needs of the growing economies of the Pacific nations have made Australia the world's leading coal exporter.

In the early 20th century, underground bituminous coal mining in Australia was the same labor-intensive, dangerous job it was in other countries. As in other countries, the Australian mining industry was plagued by labor strife. Australian mining practices and equipment lagged behind those of other nations. Even during World War II, when coal production was vitally needed for the railroads and munitions plants, appeals to the miners' patriotism were sometimes unavailing, and disruptions occurred. In 1946, the Joint Coal Board was established. The Joint Coal Board developed forecasts of coal demand and planned for ways to meet the demand. This board presided over the virtually complete mechanization of the mines in New South Wales. It often purchased modern mining machinery that mine owners could purchase or lease.

Brown Coal

Brown coal was discovered in the Latrobe Valley east of Melbourne in 1866. Use of this coal was not commercially successful for the next 50 years until methods were developed for burning it in boilers. After World War I, the state of Victoria established the State Electricity Commission and selected John Monash as its first chairman. Monash is better known as a leader of Australia's forces in World War I. Under Monash's leadership, the State Electricity Commission planned a model mining community at Yallorn, developed a strip mine for extracting the coal, built an electric power station, installed the necessary lines for the power to reach Melbourne, and set up a factory to make brown coal briquettes for heating. The Victorian brown coal industry has enjoyed steady growth ever since then.

Some estimates suggest that the Latrobe Valley contains enough brown coal to supply Victoria's energy needs for 1200 years. The seams are probably the thickest in the world, exceeding 300 feet in some places. This coal is buried under only about 30 feet of dirt and rock, so the seam is easily exposed for mining. Like other brown coal, the Australian variety has very high moisture contents (in excess of 60%) and correspondingly low heating values if burned essentially as mined. However, the State Electricity Commission has established a foresighted program of mapping and characterizing the coal. Test holes are bored at 400-meter (approximately 0.25-mile) intervals in a gridlike pattern throughout the valley. Analysis of the coal cores indicates the quality of coal that will be burned several years from now. If researchers and engineers find significant changes in coal quality, they have time to develop possible solutions to problems years before power-station boilers receive the coal. The State Electricity Commission's strategy could profitably be adopted elsewhere.

Great Britain

Coal is one of the few mineral resources that Great Britain possesses. Essentially all of the coal in Great Britain is bituminous coal and anthracite. Only one small deposit of lignite occurs in Great Britain. Lignite was recently discovered in Northern Ireland. Coal is found in the north and west of England, above an imaginary line that can be drawn from Bristol northeastward through Coventry and Leicester to The Wash. In Scotland the principal coal deposits are in the middle of the country, in the region between the northern Highlands and the rolling upland terrain in the south. Welsh coal is mined in the south of Wales, in regions centered around Cardiff and Swansea.

Great Britain could fit comfortably inside the state of Oregon. Although most regions of Great Britain are densely populated, the total population has long been smaller than the principal nations on the Continent. Nevertheless, for centuries Great Britain played a role in world affairs far out of proportion to its size and population. Similarly, the development of coal mining and the coal industry in Great Britain has had more of an impact on world history and economy than might be expected from Great Britain's 1% share of world coal resources. The deposits of coking coal near deposits of limestone and iron ore led to the development of the coke-fueled blast furnace for producing iron. The need to pump water from deep coal mines led, through a series of inventions and improvements, to the steam engine. The steam engine and inexpensive iron and steel so radically changed the nature of society that their coming sparked what we now call the Industrial Revolution. Great Britain led the way in the Industrial Revolution and for years was the most prosperous nation on earth. As early as 1800, Great Britain was mining 10 million tons of coal per year. For much of the 19th century, Great Britain was the world's leading coal producer until that role was eventually overtaken by the United States.

The high point of the British coal industry occurred in the decade before World War I. Some postwar recovery accompanied the general prosperity of the 1920s, but the industry was hit hard by the Great Depression. Coal, like the railways and some other industries in Great Britain, was nationalized shortly after World War II. The National Coal Board was created in July 1946, and the industry was nationalized on January 1, 1947. Even in the early 1950s, coal accounted for about 90% of Great Britain's power, but since then its use has steadily declined. Now only about one-third of Great Britain's power is generated from coal.

The decline of the British coal industry is not unlike that experienced in Germany. Oil became available at favorable prices and in increasing quantities. Some of the coal mines were operating at uneconomical depths or thicknesses of coal seams. Some collieries in England were already at depths below 4000 feet in the 1960s and were extracting coal from seams only 12 inches thick. The policy of the National Coal Board has been to try to stabilize production and close many of the small mines that are uneconomical to operate. The attempts to close some of the mines have led to very bitter and persistent labor disputes.

Other Nations

At least 30 nations besides those discussed possess significant coal deposits, yet each nation's production amounts to less than 1% of the world's total production. Coal is an important fuel in India, it provides Poland with much-needed exports, and it produces much of the gasoline and petroleumlike products used in South Africa. As more nations of the Third World become industrialized and explore and then develop their internal resources, their previously unknown coal deposits will become prominent. Indonesia, for example, possesses a coal that is less than 1% ash as mined and comprises about 90–95% of the reactive macerals[8] exinite and vitrinite. If sufficient supplies exist, this coal will surely find a ready market. Much is yet to be learned about the coal resources of South America. As world economic development continues, patterns of coal consumption and sales will shift, too. The lead in coal exports has moved from Great Britain to the United States and now to Australia. Coal will certainly be an important world energy source for decades to come, and undoubtedly other nations will someday come to the forefront.

Notes

[1] Berkowitz, N. *An Introduction to Coal Technology*; Academic: New York, 1979; p 18.

[2] *World Book Encyclopedia*; World Book–Childcraft International: Chicago, 1979; Vol. 4, p 582.

[3] Edgar, T. F.; Kaiser, W. R. In *The Chemistry of Low-Rank Coals*; Schobert, H. H., Ed.; ACS Symposium Series 264; American Chemical Society: Washington, DC, 1984; pp 53–76.

[4] Lerner, L. *Miracle at Springhill*; Holt, Rinehart & Winston: New York, 1960.

[5] Manchester, W. *The Arms of Krupp*; Little, Brown & Co.: Boston, 1968; p 14.

[6] Shirer, W. L. *The Collapse of the Third Republic*; Simon & Schuster: New York, 1969; p 168.

[7] AG is the abbreviation for Aktiengesellschaft, the German equivalent of incorporated. The consolidation of the major coal companies eased the planning and control of coal production and the diversification of the affected companies. Today, coal from the Saar and Ruhr is a major contributor to the resources of the European Economic Community.

[8] To be discussed in Chapter 4.

Chapter 4

Chemistry and Physics of Coal

A visual examination of coal shows that it encompasses a range of materials, from moist, crumbly brown coal to hard, glossy, black anthracite. Gertrude Stein's famous remark[1] that "a rose is a rose is a rose" could never be applied to coal. The variability of coal arises because different kinds of plants predominated in the geological periods when coal deposition was most active. Each type of plant had its own set of plant components contributing to the structure, internal energy mechanisms, and specialized activities of the plant. Also, the different plant components may have experienced varying degrees of decomposition during the biochemical phase of coal formation. Furthermore, the deposits of newly formed coal, which already incorporate different amounts of biochemical degradation acting on different types of plant components, may have experienced different degrees of metamorphism during the geochemical phase of coal formation.

Many of the properties of coal vary consistently with rank, and in some cases relationships among several properties can be discerned. Nevertheless, coal science, which might be described as the discipline that studies the geochemical history, chemical composition, physical properties, and fundamental reactivity of coal, is very much in its infancy compared with disciplines such as biochemistry or materials science. Coal is indeed a complex, diverse, heterogeneous substance; however, its composition and properties must ultimately be described in terms of the basic concepts of chemical bonding and structure. Developing such a description is by no means simple and is certainly not without controversy among coal scientists. Still, as research groups around the world uncover new information about coal, our understanding of its origins, composition, and behavior in chemical reactions is becoming clearer.

Petrography: The Components of Coal

In many research areas, including coal science, the new instruments that provide increasingly powerful probes of composition or properties are being applied. As we increase our ability to study substances in greater detail, we often forget that a simple test—a visual examination—can provide much useful information.

Lithotypes

If we look at samples of bituminous coal, even without a magnifying glass or microscope, we can distinguish bands of material running through the coal. Four types of bands are found:

1. vitrain, a glossy black material that breaks in a circular pattern (called *conchoidal fracture*) similar to the way glass breaks when chipped;
2. clarain, a glossy black material that breaks horizontally;
3. durain, a dull black material having a granular appearance; and
4. fusain, a material that resembles charcoal.

Each of these bands is a *lithotype*. Coal lithotypes are believed to arise from different conditions that existed during the deposition of the coal. For example, fusain looks like charcoal because, in a sense, it once was charcoal: Scientists believe that fusain may have formed as a result of fires sweeping through the swamp or bog in which the coal was deposited. The trees were partially carbonized or "charcoalized" by the fire, and this charcoal structure persisted into the coal that eventually formed.

Macerals

Lithotypes are composed of *macerals*, which can be distinguished under a microscope. The various macerals are grouped into categories called *maceral groups*. The three maceral groups are vitrinite, exinite, and inertinite.[2] These three groups all end in "inite" as do the names of the various macerals that make up these groups. Thus, all maceral groups and macerals end in "inite", just as all lithotypes end in "ain".

The macerals represent different kinds of coalified plant material. The decomposition of lignin, which gave rigidity to the plant tissues, results in a material that through coalification produces the vitrinite maceral group. The exinite maceral group comes from coalification of some of the specialized components of plants: Spores eventually become sporinite, resins become resinite, and leaf cuticles become cutinite. The carbonized wood or charcoal becomes fusinite, which is a member of the inertinite maceral group.

The different molecular components of plants produce the different macerals and maceral groups. Because the different plant components have different molecular structures, most likely their chemical behavior will be different. For example, the waxes present in a leaf cuticle consist of long chains of carbon atoms linked somewhere near the middle by an ester functional group. On the other hand, a resin may consist of several rings of carbon atoms joined together, it may contain no esters, and it may have less hydrogen in proportion to carbon than the wax does. A wax and a resin probably will dissolve in different solvents and behave differently when heated. To the extent that the structures of the plant components persist through coalification, the chemical structures in the different macerals will be different. Again, a different chemical structure implies different chemical behavior. Therefore, the chemistry or the behavior of a coal sample can be inferred from the proportions of macerals in the sample. This examination is called *petrographic analysis* and is especially important for assessing coal quality for coke production.

Different aggregations of maceral groups result in different lithotypes. Vitrinite, with or without small amounts of exinite and inertinite, composes vitrain. Durain also contains all three maceral groups, but in approximately equal proportions. The inertinite group macerals are predominant in fusain. The clarain lithotype appears to be a blend of vitrain and durain.

Macerals and lithotypes allow us to trace the sequence from plant to coal. Consider, for example, a cypress tree growing in a swamp during the late Cretaceous period, 70 million years ago. The components of that tree are classes of materials that create and maintain its structure, provide food and energy, and perform specialized functions or services such as sealing small wounds. Each class of plant component is made of distinct molecular species: cellulose and lignin provide structure, carbohydrates and fats serve as foods, and waxes and resins do special jobs. Macerals arise from the molecular species of the plant components, and most likely their chemical composition is related to the components. The maceral groups make up the lithotypes, and the lithotypes form the coal. Conceptually, the tree can be dissected into classes of components, and then the classes can be dissected into specific molecular components. At this level we can envision the transformation of plant components to coal components, that is, the macerals. We could arrive at the same point by starting with coal, dissecting it to the classes of components called lithotypes, and then further dissecting the lithotypes to macerals. This conceptual relationship is illustrated in Figure 4.1.

Probing the Chemistry of Coal

Chemists have probably tried every standard reaction in organic chemistry on coal. As a material for laboratory investigations, coal offers a lot of disadvantages. Any given sample is likely to be heterogeneous, combining several macerals, mineral matter, and some moisture. Coal does not dissolve completely in any solvent; thus, techniques that work best on materials in solution cannot be used. The lack of transparency of coal hinders some spectroscopic studies. Also, coal is not completely crystalline, so studies with X-rays are difficult to interpret. Coal does not melt and cannot be distilled, and heating coal causes its decomposition. An approach frequently used to study coal is to try reactions that are reasonably well-understood for simpler organic compounds. Analyzing the products of these reactions provides data that allow chemists to infer how these products might have been linked together in the coal.

Oxidation

Oxidation is one of the most popular techniques for probing the chemistry of coal. Some of the more common oxidizing agents are nitric acid, potassium permanganate, and hydrogen peroxide. The reactions of these oxidizing agents with relatively simple organic compounds are well-known or can be easily studied. For example, when toluene is oxidized with potassium permanganate, the product is benzoic acid. Simple organic compounds that are thought to have chemical structures similar to coal and that are used to test reactions are called *model compounds.* Many coal researchers spend much of their time studying the chemistry of model compounds as well as the chemistry of coal.

In principle, if benzoic acid is found as an oxidation product of coal, then part of the carbon and hydrogen in the coal might be present in molecular structures similar to toluene. Unfortunately, interpreting chemical studies of coal is not this straightforward for several reasons:

1. An investigation of several model compounds, for example, toluene, ethylbenzene, and *n*-propylbenzene, would show that they all form benzoic acid when oxidized with potassium permanganate. Thus, the compounds destroyed in the experiment cannot be absolutely determined.

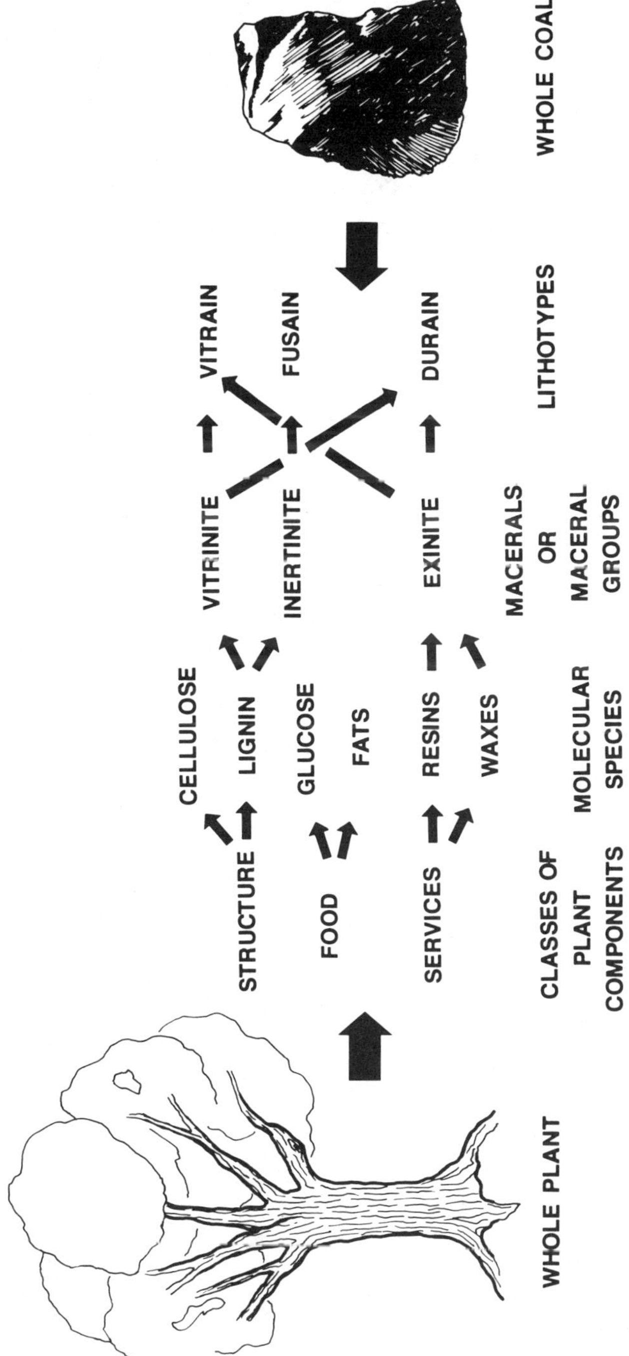

Figure 4.1. Plant components such as cellulose and lignin become the macerals in coal. Assemblages of macerals are the lithotypes that compose a lump or seam of coal.

2. Ethylbenzene contains eight carbon atoms, but its oxidation product, benzoic acid, contains only seven. The missing carbon atom escaped from the reaction as gaseous carbon dioxide. This result tells us that an analysis of the oxidation products found in the flask at the end of the experiment probably will not account for all of the carbon in the original sample. This situation is especially true for coal oxidation.
3. The oxidation of coal does not yield a single product, as in the case of toluene or ethylbenzene. Instead, the oxidation products are usually a very complex mixture of many acids. Data obtained from a careful analysis of this mixture could be used to provide a picture of the structural arrangements in coal. However, another analysis using the same data could create an entirely different picture but one equally consistent with the data.

This discussion is not meant to imply that chemical reactions of coal are useless for providing information about its composition and molecular structure. Until instrumental methods were developed for obtaining chemical information from solid coal samples in the late 1960s and 1970s, chemical reactions were about the only tool that chemists had for studying coal. Although the results of these reactions gave only a partial picture and were subject to more than one possible interpretation, they provided an idea of how some of the atoms in coal must be arranged.

Chlorination

Oxidation is not the only chemical approach used to study coal. About 100 years ago, chlorine was found to react vigorously with coal. For some coals the product of this reaction could dissolve completely in alkaline solutions. Bituminous coals and lignites can be chlorinated to make solids that look like resins and dissolve in alcohol or other organic solvents. This reaction has not been studied extensively, perhaps partly because no commercial interest in the products exists.

Extraction

The extraction of coal with organic solvents provides a solution of coal-like material that, in principle, could be studied further by spectroscopy or chromatography. The problem is that in most cases the extracts seem to be about as complicated as the original coal. A study of coal extracts has not advanced our understanding of coal chemistry very far. To be sure, a careful selection of solvents can remove specific classes of compounds from coal.

For example, the extraction of low-rank coals with benzene yields waxes, which are organic acids, alcohols, or their esters containing 20–60 carbon atoms.[3] The yields of specific classes of compounds, such as the waxes, are often small. Although characterization of the components of such extracts may be easier, the components represent only a small portion of the original coal. Thus, only a small portion of the coal is being characterized.

Effects of Heat on Coal

Volatile Matter

Heating coal in an inert atmosphere removes material called *volatile matter.* The volatile matter consists of a mixture of gases, low-boiling-point organic compounds that condense to an oily material, and heavier organic compounds that condense to tar. The generation of volatile matter is a result of *pyrolysis*, which means splitting by heat or fire. Most of the materials in the volatile matter are liberated by breaking chemical bonds in the coal during the heating process.

The products of pyrolysis, like those of oxidation, are very complex mixtures. In some pyrolysis experiments, we cannot be sure that the original coal components liberated by pyrolysis did not further pyrolyze and thus doubly complicate product analysis. This situation led to the remark, attributed to the great German coal scientist Franz Fischer, "The examination of tar reflects the composition of a coal in about the same proportion as the smoke from a burning library would reflect the nature of the contents of the burning volumes."[4]

Caking Coals

When most bituminous coals are heated, they not only liberate volatile matter but soften to a plasticlike mass that swells and resolidifies into a hard, porous solid. This behavior is typical of high-volatile through low-volatile bituminous coals (Table 1.1, p 13) but is especially noticeable for the medium-volatile bituminous coals. Coals that show this behavior are called *caking coals.* Some caking coals make a strong solid product that is used as coke in blast furnaces and other metallurgical operations. These strongly caking coals are the *coking coals.* The similarity of names is unfortunate, and they are sometimes used interchangeably. Strictly speaking, all coking coals are caking coals, but a caking coal is not necessarily a coking coal.

In the United States, caking behavior is measured by the *free-swelling index.* This test involves heating a coal sample under standardized conditions, comparing the resulting solid mass with a series of standardized outlines, and then assigning an index between 1 and 9. Outlines for three of the indexes are illustrated in Figure 4.2. A free-swelling index lower than 3 is an indication that the coal is weakly caking. The most important application of the free-swelling index is in selecting coals for making coke to be used in the steel industry. The petrographic composition and free-swelling index give the plant operator the information needed to select and blend coals to produce high-quality coke. Exinites show great fluidity, whereas inertinites exhibit no plastic behavior at all. Vitrinites are intermediate between these two maceral groups.

The first effect of temperature on the caking process appears to be the breaking of enough bonds in the coal to create a liquid, called the *primary tar,* which acts to increase the mobility of coal fragments within the plastic mass. After further heating, the primary tar breaks down into gases or vapors that bubble through the plastic. The release of the volatile products from the primary tar is probably what causes the swelling. Continued heating causes the almost complete destruction of the primary tar, but because the primary tar was responsible for the fluidity, its destruction results in resolidification of the nonswollen material. On a molecular scale, aromatic rings in the coal begin to fuse into structures that begin to approach, on a small scale, the structure of graphite.

Instrumental Probes

Chemists have examined coal by reacting it with various reagents or breaking it apart with heat since before the beginning of the 20th century. Not much progress was made in determining the molecular structural features of coal for the reasons we have discussed: difficulties in accounting for all the matter in coal, inability to ascertain that the products of experiments were not altered more than intended, and lack of an unequivocal way to conceptually reassemble the products of experiments into an idea of a structure. After World War II, chemists increasingly began

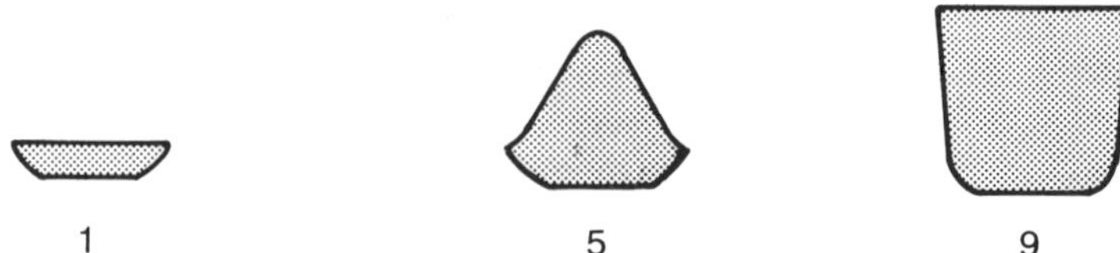

Figure 4.2. After a free-swelling index experiment, the size and shape of the remaining solid are used to assign a numerical value of the index. A coal that has a free-swelling index of 1 has no caking properties, whereas a coal that has an index of 9 is highly caking. Intermediate values represent intermediate caking tendencies. A free-swelling index of 5 indicates a candidate for coke manufacture.

to apply instrumental techniques to the study of coal. These techniques do not require that coal be reacted or degraded and therefore eliminate the uncertainty of changes caused by chemical processes. Unfortunately, instrumental methods cannot account for all the matter in coal and cannot conceptually reassemble the products of experiments into an idea of a structure.

X-ray Diffraction

The first instrumental technique used in coal analysis was X-ray diffraction. When a beam of X-rays is focused on a material, the electrons surrounding each of the atoms in the material "bend" the X-ray beam, causing it to flare into a pattern. This phenomenon is called *diffraction*. An example of the diffraction of light can be seen by looking at a distant light source between two fingers held close together; the fringe of dark strips appearing between the fingers is caused by the diffraction of light passing through the narrow slit between the fingers. X-rays diffracted from a crystal can be detected by photographic film or electronic detectors. Figure 4.3 shows a schematic diagram of an X-ray diffraction experiment using a powdered coal sample. The geometric pattern of diffracted X-rays and the intensity of the X-ray beam at each point in the diffraction pattern provide the information that is needed to calculate the locations of and distances between the atoms in the crystal.

X-ray diffraction has become an extremely powerful tool for studying the molecular structures of crystalline substances. It has provided structural information on even very complex materials such as DNA. Unfortunately, coals are not entirely crystalline. Deciphering the X-ray data from an amorphous (i.e., noncrystalline) material is not an easy, straightforward procedure. Like chemical tests, X-ray diffraction does not give an unequivocal structure for coal. Nevertheless, X-ray studies have added to the understanding of the atomic arrangements in coal.

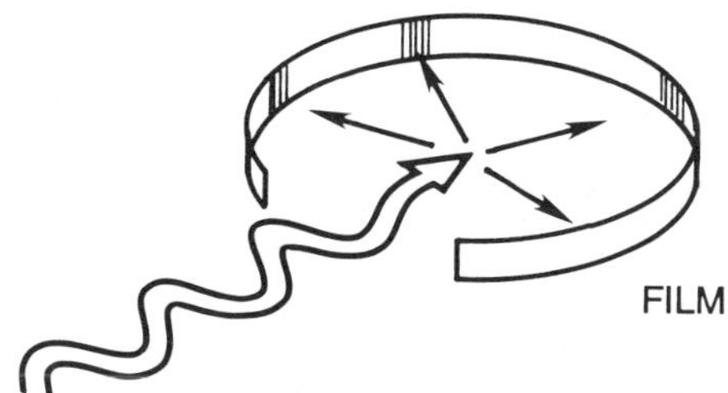

Figure 4.3. In an X-ray experiment with a powdered sample, the incoming beam of X-rays can be diffracted from all of the planes of atoms in the sample. The diffracted X-rays produce a pattern of lines on the film surrounding the sample. The pattern of lines is characteristic of the structure of the sample.

In particular, the X-ray studies have shown that for most coals, except anthracites, 70–90% of the carbon is arranged in clusters of benzene rings fused together, usually from two to four rings per cluster. Comparing a series of bituminous coals of increasing carbon content shows that at the higher carbon contents, the small clusters of rings begin to stack atop each other in a miniature version of the graphite structure. Virtually all the carbon in anthracites is contained in ring systems; the structures are much larger than they are in bituminous coals and may consist of large sheets of tens of benzene rings fused together.

Infrared Spectroscopy

The second instrumental technique used in coal analysis is infrared spectroscopy. Infrared radiation is a portion of the total spectrum of electromagnetic radiation that runs from the very short wavelength gamma (γ) rays to the very long wavelength radio waves. Infrared radiation has wavelengths longer than red light, so it is not visible to our eyes. The vibration or rotation of two atoms in a chemical bond can be stimulated by infrared radiation. Vibrations of carbon–hydrogen and carbon–carbon bonds are shown in Figure 4.4. Physically, this effect is observed by an absorption of infrared radiation at a wavelength characteristic of the energy required for a particular vibration or rotation. In the laboratory, an infrared study is done by scanning the spectrum of infrared wavelengths and recording the wavelengths at which the substance being tested absorbs the radiation, as illustrated in Figure 4.5. This practice is called *infrared spectroscopy.* For a particular pair of atoms bonded in a particular molecular configuration, such as the bond between carbon and hydrogen in a benzene ring, the wavelength at which the infrared radiation is absorbed is characteristic of

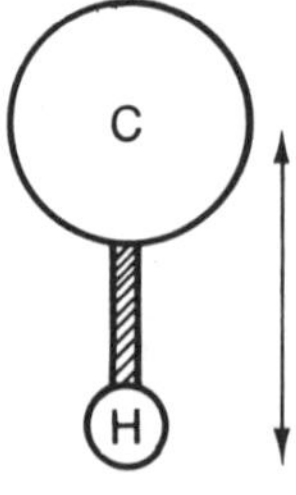

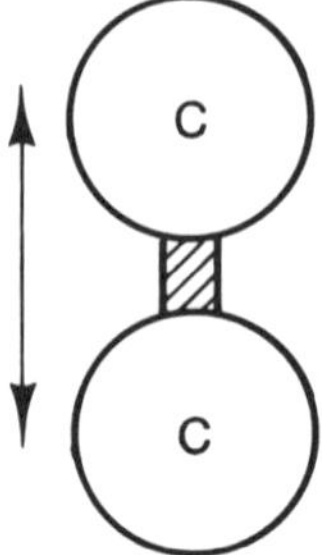

Figure 4.4. The vibrations of chemical bonds can be visualized as two masses connected by a spring. A hydrogen atom has one-twelfth of the mass of a carbon atom. A hydrogen–carbon bond can vibrate at a much higher frequency than a carbon–carbon bond. In experiments using infrared spectroscopy, the hydrogen–carbon bonds absorb radiation at higher frequencies (shorter wavelengths) than the carbon–carbon bonds.

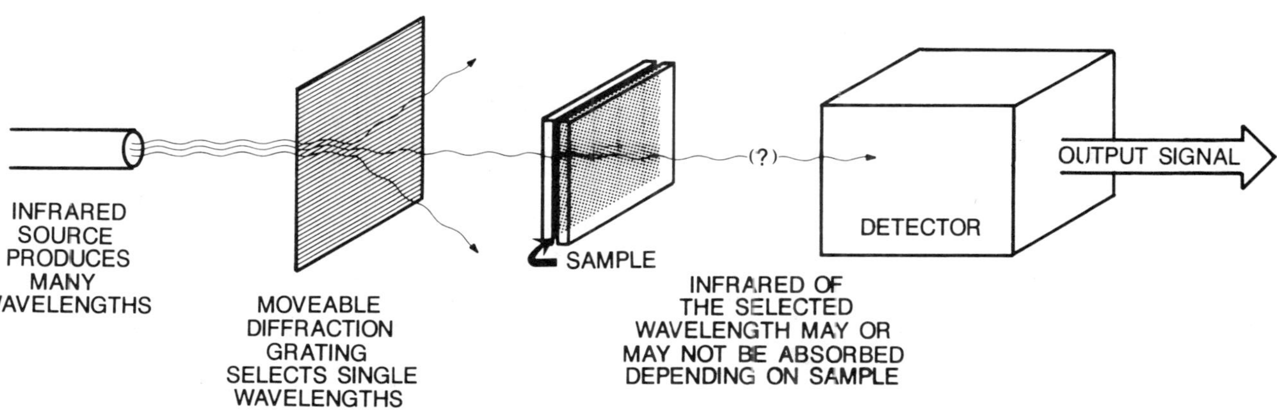

Figure 4.5. In an experiment using infrared spectroscopy, selected wavelengths of infrared radiation are passed through the sample. Whether or not the radiation of a given wavelength is absorbed depends on the structural arrangement of the atoms in the sample. The output from the instruments indicates which wavelengths have been absorbed. This information allows the chemist to deduce information about the structure of the sample.

that bond and is essentially the same for every substance containing that bond. Knowing the assignments of particular wavelengths to particular bonds, a chemist studying a new substance can record the pattern of absorption from the infrared spectrum and deduce information about the molecular structure. For materials that have relatively simple structures, the elemental analysis, molecular weight, and infrared spectroscopy data may be adequate to deduce the structure unequivocally. For more complex materials, supporting data from other instrumental methods or wet chemical analyses are needed. For very complex materials, such as coal, which do not have an exact structure, infrared spectroscopy still provides information on the kinds of bonds present.

Infrared spectroscopy shows that in coal, carbon and hydrogen are present in both aromatic and aliphatic rings and in aliphatic chains. Double or triple bonds between carbon atoms are very rare. Oxygen is contained in phenolic structures and as ethers linking two ring systems together. In lower rank coals, below about 80% carbon, oxygen also occurs as organic acid groups and as *methoxyl* groups, a specific type of ether structure that occurs in lignin.

Nuclear Magnetic Resonance Spectroscopy

Both X-ray and infrared evidence show that the carbon in coal is present in both aromatic and aliphatic forms. For most chemical reactions, the chemistry of aliphatic carbon is different from that of aromatic carbon. In studying coal chemistry, knowledge of the fraction of the total amount of carbon present in each form is helpful. The fraction of carbon present in aromatic structures is called the *aromaticity*. X-ray and infrared data have been used to deduce the aromaticity of coal. Measurement of aromaticity is also a specialty of the third major instrumental technique in coal chemistry, nuclear magnetic resonance spectroscopy.

The physical basis of nuclear magnetic resonance depends on the tendency of protons and neutrons in the atomic nucleus to pair together. When an atomic nucleus contains an unpaired proton or neutron, such a nucleus spins and acts like a tiny magnet. If this tiny nuclear magnet is surrounded by a magnetic field from outside, the nuclear magnet will align itself to be parallel or antiparallel to the external field. The parallel alignment is the more stable. Thus, energy would have to be absorbed to flip the nuclear magnet into the antiparallel alignment; if an antiparallel nucleus flipped back to the parallel alignment, energy would be released. The energy associated with these transitions is in the radio-frequency region of the electromagnetic spectrum. An experiment would consist of placing a sample in a magnetic field, exposing it to radio-frequency energy to flip the nuclear magnets into the antiparallel alignment, and then observing the energy emitted by the sample when the magnets flip back to the parallel alignment. This experiment is illustrated schematically in Figure 4.6, and an

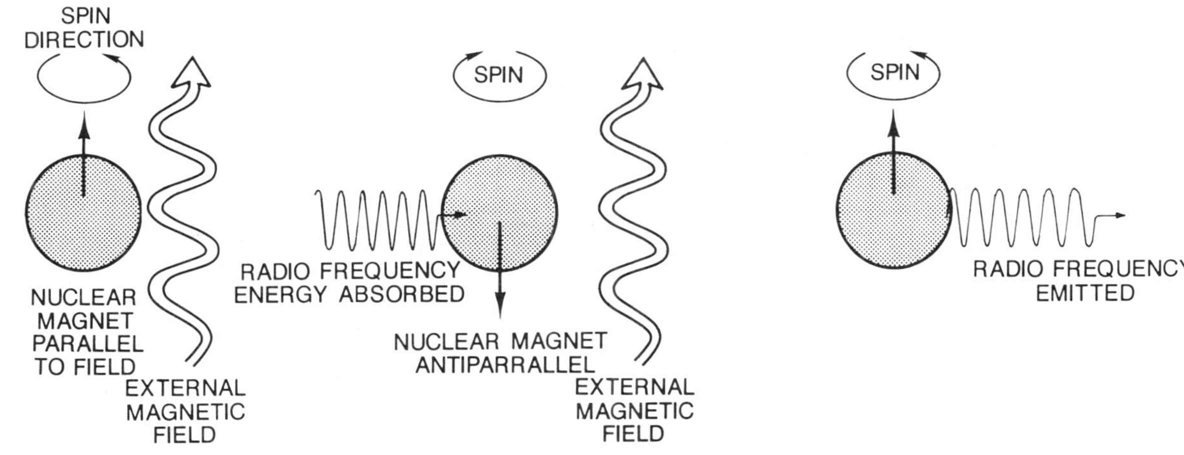

Figure 4.6. In one version of the nuclear magnetic resonance experiment, the nuclear magnet is aligned parallel to an externally applied field. The absorption of radio-frequency energy flips the magnet to antiparallel alignment. The magnet then relaxes back to the parallel alignment and emits energy as it does so. The emitted energy is detected as a signal by the instrument.

experiment for measuring the nuclear magnetic resonance spectrum is shown in Figure 4.7. The external magnetic field experienced by a nucleus is not exactly the field of the magnet used in the experiment because a particular nuclear magnet will be shielded from the external field by whatever cloud of electrons exists in the molecular environment around that nucleus. Because the nuclei in a sample will "see" slightly different magnetic fields as a result of the way they are shielded from the external field, the energy absorbed or emitted when the nuclear magnets flip will also be slightly different, depending again on the molecular environment of the nucleus. Therefore, an element having a nucleus with spin will provide a series of responses in a nuclear magnetic resonance experiment; each signal will arise from specific molecular configurations for some of the atoms of that element. As is the case with infrared spectroscopy, nuclear magnetic resonance experiments with model compounds of known structure provide the information needed to relate a specific signal to a specific structure.

An infrared experiment will detect the vibration or rotation of almost any covalent bond regardless of the types of atoms forming the bond. In contrast, nuclear magnetic resonance is restricted only to atoms having nuclear spins. Different atoms respond in such different regions of the radio-frequency spectrum that one particular nuclear magnetic resonance

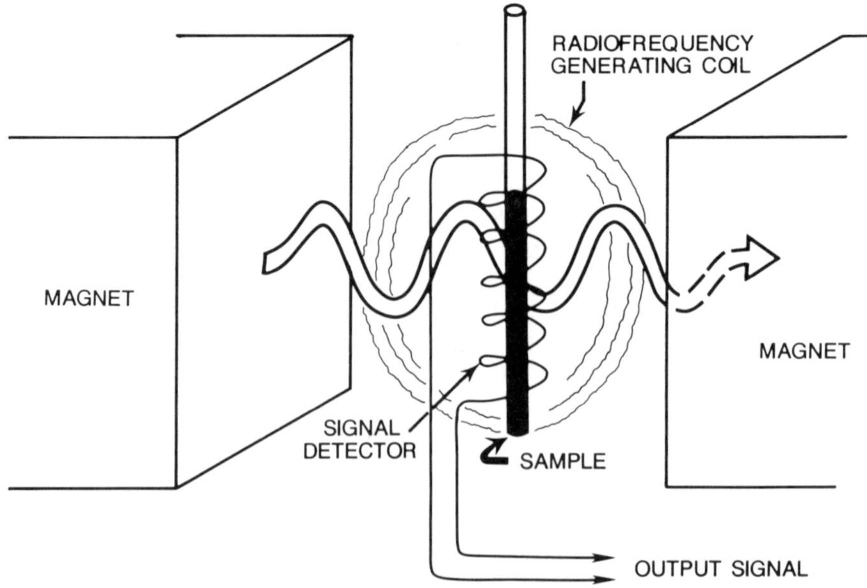

Figure 4.7. In the nuclear magnetic resonance experiment, the sample is contained in an external magnetic field. The sample receives radio-frequency energy that is applied perpendicularly to the magnetic field. The energy emitted when the nuclear magnets relax to an alignment parallel with the field is detected and produces an output signal from the instrument.

experiment will only provide data on atoms of one element. Fortunately, the ordinary isotope of hydrogen contains only a single proton in its nucleus. Because the proton has no neutron with which to pair, the nucleus of hydrogen must have a spin and therefore is responsive in nuclear magnetic resonance experiments. This situation allows this technique to provide much information about the arrangement of hydrogen atoms in organic compounds. The ordinary isotope of carbon, carbon-12, has a nucleus containing six protons and six neutrons. The protons and neutrons are paired, leaving no extra spin to make a nuclear magnet. However, the isotope carbon-13 has one "extra" neutron and thus can be studied by nuclear magnetic resonance. Although carbon-13 is much less abundant than carbon-12 (for every 100 atoms of carbon-12 in a sample, only one atom of carbon-13 is found), enough is present for detection.

A nuclear magnetic resonance experiment usually works best if the sample is in solution. The easy mobility of dissolved molecules lets the nuclear magnets align with the external field. The signals produced in the experiment are sharp responses that can be assigned to structural features known to exist in model compounds. The early work on nuclear magnetic resonance of solid samples was disappointing. Mobility in a solid is much more restricted than mobility in a liquid. Instead of giving sharp responses, the solid samples produced smeared signals that were difficult to interpret. Because coal is not completely soluble in any solvent, the first nuclear magnetic resonance studies of coal were of hydrogen in solutions of coal extracts, pyrolysis products, and products of other kinds of processing, such as liquefaction. Although this work was unquestionably useful in understanding the structures of those materials, relating the results to the structure of coal was difficult.

Recently, advances in experimental techniques and in ways of manipulating the data have made possible carbon-13 nuclear magnetic resonance studies of solid coal. This method of characterizing coal is an active area of research in laboratories around the world. The results show that aromaticity increases as rank increases and ranges from about 0.6 in lignites to 0.9 in low-volatile bituminous coals. (An aromaticity value of 0.6, for example, means that 60 of every 100 carbon atoms are present in aromatic structures.) The aromaticity of anthracite is almost 1.

Structure of Coal

Chemical probes and instrumental methods provide information that can be assembled to give a composite picture of the structural relationships in coal. Coal researchers concur that coal does not have a structure in the sense either of a polymer or a large biological macromolecule. A polymer, such as polyethylene or polystyrene, consists of large molecules, often having molecular weights in the tens or hundreds of thousands. A polymer has a simple repetitive structure; if we looked at a portion under a mythical microscope capable of showing us the individual atoms and then looked at a second portion several hundred carbon atoms away, the two portions would look very much alike. A biological macromolecule, such as DNA, may have a very complex structure, but one that can be determined in detail at least in principle. Once we know the structure, we can be confident that a second sample from the same source is essentially identical. In the case of coal, if the exact structure of a particular sample was determined, most likely another piece of coal with the same structure would never be found, even from coals of the same rank or from the same mine. Thus, why should we bother with structure determinations?

When a coal scientist talks about the structure of coal or draws a structural formula to represent the structure of coal, that formula actually represents some sort of statistical or average structure that happens to agree with the ultimate analysis, aromaticity, amounts and types of functional groups, and major carbon structures. Are these structures useful? Consider this analogy: Current theories of the electronic structure of atoms tell us that atoms do not look like the neat "solar system" atoms sometimes presented in elementary textbooks. Electrons are not dots arrayed around an atom in neat circles. However, if we had to explain how an atom of sodium reacts with an atom of chlorine to make sodium chloride, the easiest way would probably be to move the electron dot from the outermost circle of the sodium atom over to the last position in the outer circle of the chlorine atom. Similarly, although the structure of a piece of coal may not exactly mimic the average structural formula that someone has drawn, such structural formulas still provide good models to illustrate what happens when coal undergoes reactions.[5] These structures allow chemists to think about the sites in coal where reactions might occur, especially reactions that could break the structure apart, and to think about the chemical nature of the products of such reactions.

Ultimate Analysis

The starting point for assembling data to develop a picture of coal structure is the ultimate analysis. We know from results of thousands of ultimate analyses that carbon content varies in a regular way with rank, from less than 70% in brown coal to about 95% in anthracite. Until carbon content reaches about 90%, the hydrogen content is fairly constant at about 5% ± 0.5%. Higher than 90% carbon, the amount of hydrogen drops off rapidly as the amount of carbon increases, falling to about 2% in anthracite. Oxygen content decreases steadily as rank increases, ranging from 25% in lignite to 1% in anthracite. Nitrogen is a minor constituent and tends to compose about 1% of almost all coals regardless of rank. The total sulfur in coal is quite variable, but sulfur analyses indicate that the amount of organic sulfur is usually less than 1.5% and does not vary much with rank. The two components that show the greatest variability with rank are carbon and oxygen.

X-ray Diffraction

X-ray diffraction shows that coal tends to become more like graphite as the carbon content increases. The number of aromatic rings fused together increases as rank increases and ranges from 1 or 2 in lignites, to 3–5 in bituminous coals, to more than 10 in anthracites. At carbon contents above 85%, the ring clusters begin to show some stacking or vertical arrangement. As coal becomes more graphitelike, the aromaticity probably increases. Indeed, nuclear magnetic resonance shows that aromaticity rises from about 0.6 in lignites to 0.95 in anthracites.

Infrared Spectroscopy

The types of oxygen structures can be determined by infrared spectroscopy or by various chemical reactions. The principal oxygen structures are ethers, phenols, carbonyl groups (primarily ketones), and carboxylic acids. Not only does the total oxygen content decrease as rank increases, the oxygen groups also vary with rank. Methyl ethers, or methoxyl groups, occur only in low-rank coals containing less than 75% carbon. The carboxylic acids occur only in coals containing less than 80% carbon. Phenol and carbonyl structures persist in coals up to about 90% carbon. Methoxyl, phenol, and carboxylic acid groups are all attached to rings. Because these groups vanish, in a sense, from coals of increasing carbon content, aromatic rings should be expected to have fewer structures attached to them as carbon content or rank increases. This expectation is confirmed by infrared or nuclear magnetic resonance studies.

Thus, as rank increases, more of the carbon is incorporated in aromatic ring systems, and the systems themselves gradually get larger. Oxygen content, on the other hand, decreases as rank increases; the oxygen structures that occur as substituents attached to rings become less prevalent.

Bridging of Aromatic Rings

The hooking together or bridging of ring systems is most likely brought about by the aliphatic carbon atoms or by ether oxygens. The aliphatic carbons can occur in two types of structures: chains or aliphatic rings. Because coal is relatively hard, does not melt, and does not dissolve completely, the three-dimensional arrangement of linkages between ring systems is probably very much like a cross-linked polymer. (Natural rubber is an example of a linear polymer, consisting of 1000–5000 structural units linked end-to-end; in contrast, in a cross-linked polymer, the sequence of structural units is linked together in two or three dimensions. The ion-exchange resins used in water-softening units are examples of cross-linked polymers.)

Incorporation of Nitrogen and Organic Sulfur

Not much is known about how nitrogen and organic sulfur are incorporated in the coal structure. Scientists generally surmise that the atoms of these elements are incorporated in aromatic ring systems as occasional replacements for carbon atoms.

Conclusions

The structure of coal is thus a three-dimensional cross-linked polymer consisting of aromatic ring systems linked by aliphatic carbon bridges or ether oxygen bridges. In low-rank coals, the aromatic systems are one ring or two rings fused together that contain abundant methoxyl ether, phenol, and carboxylic acid groups. As rank increases, the sizes of the ring systems gradually increase. At about 85% carbon, the rings begin to line up on top of each other and the oxygen groups become less abundant. Higher than 90% carbon, only a few bridging ether oxygens may remain. In anthracite at about 95% carbon, the ring systems incorporate tens of individual rings fused together. Graphite is, in a way, the ultimate coal; it contains 100% carbon, has an aromaticity of 1, and contains fused rings of virtually infinite extent that are neatly stacked.

Inorganic Constituents of Coal

Coal contains a rich variety of inorganic constituents. Mineral grains in coal can sometimes be seen with the unaided eye and are easily seen through a microscope. When coal is burned it leaves behind the unconsumable inorganic residue known as *ash*. The ash is the product of reactions or transformations of the inorganic components of coal caused by the high temperatures of the combustion process. Several questions concern these inorganic species: Where did they come from? What are their components? How do they behave?

Sources

Several sources contribute inorganic material to coal. The first is the coal-forming plants themselves. Most plants contain at least small amounts of inorganic substances; that is why burning logs in a fireplace always leave some ash. However, some plants concentrate significant quantities of inorganic material. For example, the scouring rushes concentrate enough silica in their cell walls to give them a gritty texture that pioneers found useful for cleaning cooking utensils or scouring floors. The scouring rushes are examples of the last surviving genus (the horsetails) of a class of plants that were prevalent in the coal swamps of the Carboniferous period. Some of the inorganic components of coal were in the original plant material.

During the diagenetic stage of coal formation, as the organic sediments were accumulating in the swamps, ample opportunities occurred for inorganic debris to be deposited by either water currents or wind. This debris would settle into the organic sediments and become part of the coal that eventually formed. Clay particles are deposited into coal in this way. Minerals transported into the coal-forming environment by wind or water are called *detrital minerals.*

Water will continue to percolate through cracks and crevices in a coal seam while the seam is forming and after the coal has formed. The water percolating through the coal can precipitate minerals in the cracks. Carbonate minerals and pyrite are deposited in coal in this way. Low-rank coals contain significant amounts of carboxylic acid groups in their structure. These acid groups react with ions dissolved in the water moving through the coal, much as ion-exchange resins remove ions from domestic water supplies. Sodium, calcium, and magnesium ions are particularly likely to become attached to acid groups in low-rank coals.

Incorporation

Some inorganic components are incorporated in coal as coordination complexes or chelates. This way of incorporating elements in coal is not well-understood (that is, from the perspective of the inorganic chemistry of coal). Some of the aluminum in low-rank coals may be present as a complexed species. A portion of the small amounts of nickel in coal may be coordinated to the organic sulfur.

Almost every element, except for the highly unstable radioactive ones and the noble gases, is present in coal at some level of concentration. These elements are incorporated in and distributed through coal in diverse ways.

Cations of sodium, magnesium, and calcium are associated with organic acid groups. In low-rank coals of the United States, almost all of the sodium and most of the magnesium and calcium are present in this form. Some potassium also occurs as exchangeable cations. Australian brown coals also contain some aluminum and iron in exchangeable forms. Because coals containing more than 80% carbon do not normally have carboxylic acid groups, this form of inorganic material is only important in brown coals, lignites, and subbituminous coals. Clay minerals are aluminosilicates, usually of potassium. Some clays may also incorporate sodium, calcium, or iron.

Clays account for much of the aluminum and some of the silicon found in coal. Clay particles are sometimes found in plant cell cavities, as shown in Figure 4.8. Quartz also contributes silicon. Iron is contained in sulfide minerals, principally pyrite and marcasite. Pyrite is also sometimes found as crystals growing in cell cavities remaining from the original plant material, as shown in Figure 4.9. Some iron and aluminum may be present as their hydrous oxides. Carbonate minerals include calcium and magnesium carbonate (dolomite). Iron and calcium sulfates may occur in some coals; however, sulfate minerals are generally unimportant.

Ash

Elemental Components. Ash is the product of heating the inorganic constituents during the burning of coal. The principal elemental components of ash are those just discussed:

- sodium,
- calcium,
- magnesium,
- potassium,
- aluminum,
- silicon,
- iron, and
- sulfur.

These eight elements, as well as titanium and phosphorus (which generally occur in minor amounts in most coals), are reported when ash is analyzed in the coal laboratory. The ash analysis results are reported as the oxides; for example, sodium is reported as the percent of sodium oxide, and iron is reported as the percent of ferric oxide. This convention is followed, although virtually none of the elements are present in ash as their respective oxides. Relationships between the behavior of ash in coal-processing operations and ash composition have been developed by establishing an empirical equation that relates some parameter of behavior to the amounts of "oxides" in the ash. Such empirical equations have sometimes successfully predicted ash behavior from ash composition. However, they provide few insights into the actual chemical basis for the observed behavior because elements generally do not occur in ash as their oxides. In addition, the temperature at which ash is traditionally produced for subsequent analysis, 1382 degrees Fahrenheit, is not a temperature at which any real coal process operates. Consequently, much current research on the inorganic chemistry of coal focuses on the characterization of the actual chemical forms of the elements

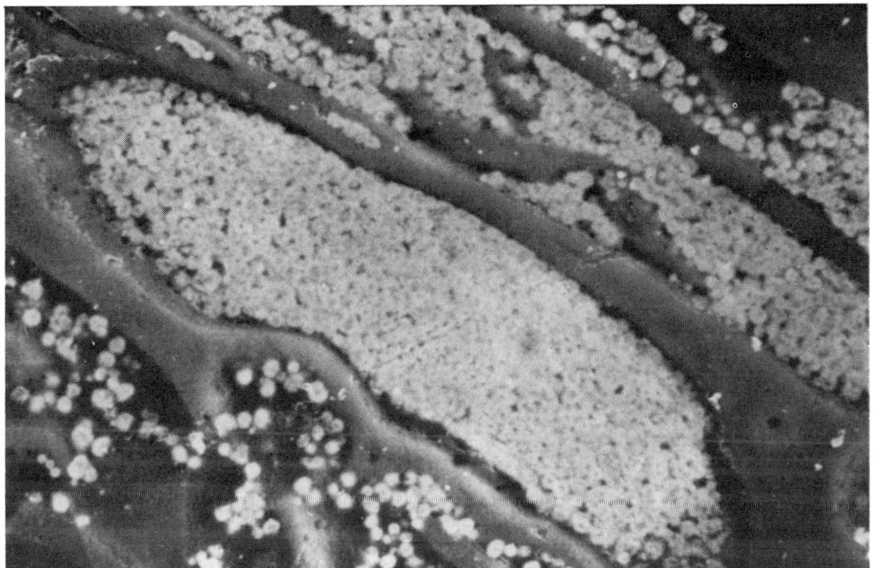

Figure 4.8. The bright material in the center of this photomicrograph is a fragment of kaolinite growing in a plant cell cavity in lignite. Kaolinite is a clay mineral that commonly occurs in coal. (Courtesy of Chris Zygarlicke.)

Figure 4.9. The dark oval structures in the center and lower right-hand corner of this photomicrograph are cell cavities in the plant material that formed this sample of lignite. The white material in the cell cavities is pyrite. The pyrite crystals are just beginning to grow in the cells. (Courtesy of David Kleesattel.)

in coal and the transformations of chemical composition and mineralogy as temperature increases.

Interest in studying the remaining inorganic elements, those that occur in coal in minor or trace amounts, arises from three general concerns:

1. The concentration or presence of some minor elements provides clues that help geochemists learn more about the deposition and subsequent geological history of the coal. For example, boron has been used as an indicator of the degree of salinity of the environment during coal deposition.
2. Arsenic, selenium, and mercury occur in trace amounts in coal and could harm the environment if they are released when the coal is used. This concern can be addressed by studying the ways in which such elements are held in the coal and what happens to them when the coal is burned. These elements need to be tracked through whatever process is using the coal to find out where they go. For example, they could be vaporized or immobilized in the ash. Then, the necessary actions must be taken to ensure that the harmful elements are not ultimately released to the environment.
3. Coal may possibly be used as a source of some rare elements. In the early 1950s, domestic sources of uranium were urgently needed for fueling nuclear reactors and building nuclear weapons. Several government agencies conducted research programs on extracting uranium from coal ash. Today, interest has shifted from nuclear fuels and weapons to solid-state electronics. Coal ash is now being considered a potential source of gallium and germanium, two elements that are ingredients of semiconductors.

Temperature Effects. To design and operate coal-processing equipment properly, the effect of temperature on the coal ash must be known. Will the ash melt and have to be handled as a molten slag? Will it soften or partially melt just enough to stick together? Or will it remain solid?

The standard procedure for determining the behavior of ash at high temperatures is the *ash fusion test.* The test begins with a sample of ash formed into a triangular pyramid. The sample is heated under standardized conditions. The analyst watches its behavior as the temperature increases. The first effect will be a slight rounding of the apex of the pyramid. The temperature at which this rounding occurs is noted and reported as the *initial deformation temperature.* Continued heating further softens and spreads the sample. At the *softening temperature*, the height of the sample is equal to the width of its base; when the height is reduced to one-half the width of the base, the *hemispherical temperature* has been reached. Finally, the sample will appear to be completely molten and spread out over its support. The temperature at this point is the *fluid temperature.* Sometimes only one temperature, the softening temperature, is reported as the result of an ash fusion determination.

The ash fusion determination was first developed to help predict whether ash would form *clinkers* (fused stony matter formed from impurities) when coal was burned in a stoker. (Clinkers were too big and hard to pass through the ash-removal system and interfered with the operation of the stoker.) Stokers are no longer used in large installations burning high tonnages of coal. Because ash fusion behavior is judged by the analyst, the analyst's experience greatly determines the results of experiments. (However, recently developed automated equipment, which incorporates a video camera to record the ash fusion events, provides a significant improvement.) The standard procedure set up by the ASTM indicates that two determinations on the same ash sample are acceptable if they agree within 50 degrees Fahrenheit. Although the original need for the ash fusion test has largely faded and the test relies on the analyst's judgment, ash fusion determinations remain a traditional part of coal characterization. Many empirical correlations have been developed to relate ash fusion data to ash behavior during coal processing. The sequence of events in an ash fusion test is illustrated in Figure 4.10.

One use of ash fusion data is to discriminate between slagging and nonslagging coals. If the softening temperature is below 2200 degrees Fahrenheit the coal is said to be a *slagging coal.* It would best be used in systems that have been designed to remove the ash as a slag. *Nonslagging coals* have ash-softening temperatures above 2600 degrees Fahrenheit. The ash of such coals would not likely melt in conventional processing equipment. A gap of 400 degrees Fahrenheit occurs between these criteria. Coals having ash-softening temperatures of 2200–2600 degrees Fahrenheit might or might not form slag, depending on the processing conditions. For these coals, the equipment should be designed and operated to prevent the ash from melting or to melt the ash and keep it flowing to be removed as slag.

Use of a slagging coal requires some knowledge of the fluid properties of the slag. The equipment operators need assurance that the slag will flow and will remain fluid to be removed from the equipment without problems. The property most intensely studied in relation to slag behavior as a liquid is *viscosity,* which measures how much the liquid resists flowing or, in other words, how thick the liquid is. "Slower than molasses in January" indicates that molasses has high viscosity at low temperatures. Because the slag will cool off as it flows from the hottest part of the combustion reaction to the point of slag removal, the variation of viscosity with temperature is an important determination. The viscosity of slag increases as temperature

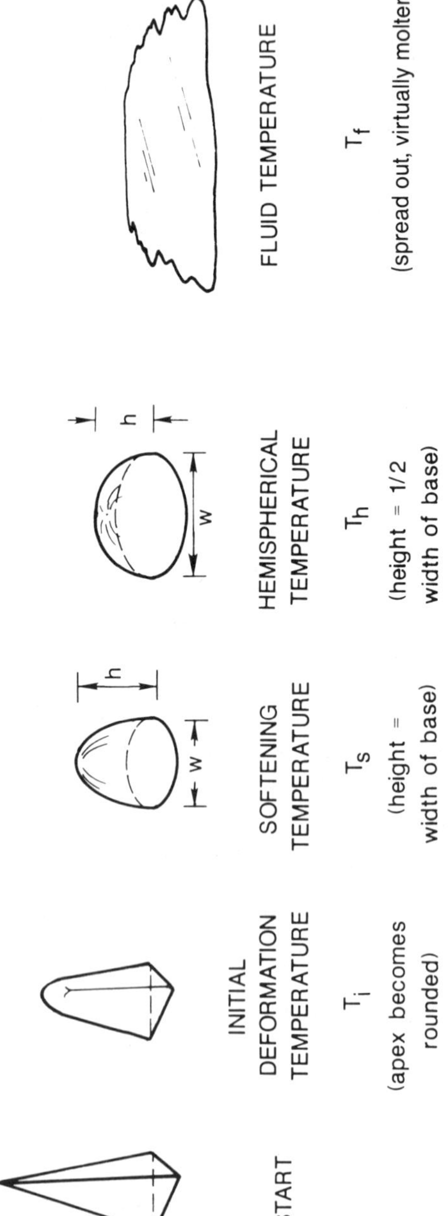

Figure 4.10. The stages of the ash fusion test, which range from a pyramidal cone of ash to a puddle of molten ash.

decreases. At high temperatures the viscosity increases only slightly as temperature decreases. Many slags reach a point at which the viscosity suddenly begins rising enormously as very small changes in temperature occur. The temperature at which this transition occurs is called the *temperature of critical viscosity.* This temperature represents the point at which the slag becomes too viscous to flow or drain from the equipment.

Physical Properties of Coal

Coal has four physical properties:

1. porosity,
2. density,
3. hardness, and
4. grindability.

These properties help determine the behavior of coal during handling and processing. Porosity also plays a role in determining the course of some reactions of coal.

Porosity

The *porosity* of a substance is the percentage of the volume of an apparently solid piece that is actually void space or pores. To determine porosity, the apparent volume of a sample, pores and all, and the volume of only the solid itself must be known. For samples having irregular shapes, volume may be difficult to measure directly, but it can be easily calculated from density measurements.[6] A porosity determination of coal involves measuring the density in mercury, which does not penetrate the pores at atmospheric pressure, and again in helium, which completely penetrates the pores. The volume calculated from the helium density is the volume of the coal, and the volume calculated from the mercury density is the volume of the coal and pores.

Interest in porosity stems from several considerations. Any process that uses coal involves a reaction of the coal with a liquid or gas at the surface of the coal. The extent of reaction will be governed by the amount of surface that the reactant can reach. The total surface of the coal is not just the surface of the exterior of the coal, but also is the surfaces making up the walls of pores in the coal. Usually, this so-called internal surface is far larger than the external surface. The size and extent of pores affect the ability of reagents to contact the coal surface and also affect the removal of reaction products so that fresh surface can be exposed for further reaction. Porosity governs the loss of moisture during coal drying and the reabsorption of moisture by the dried coal. Porosity may also relate to the strength of coal particles if the pores represent possible sites of weakness where breakage can occur.

The porosity itself is not as important as the sizes of the pores and the amount of surface area exposed because of the pores. The surface area measures the space available for reactions to occur. Pore size limits the size of molecules that can penetrate the coal and participate in reactions. A high-porosity coal having extremely small pores might not participate in reactions as easily as a lower porosity coal having larger pores.

An approach to measuring the distribution of pore sizes is to force mercury into the pores at high pressure. Although mercury does not penetrate the pores at atmospheric pressure, it does at elevated pressures. The higher the pressure, the smaller the pores the mercury will penetrate. Measuring the mercury penetration as a function of pressure makes it possible to calculate the amount of porosity attributable to the various pore sizes. For most coals, most of the porosity is contained in pores of radii smaller than 40 angstroms (about 150 billionths of an inch), including some pores ranging from 5 to 8 angstroms (about 20–30 billionths of an inch). The tiny 5–8-angstrom pores may act as connecting passages between larger pores. To put these figures in perspective, consider the nearly invisible tungsten-wire filament in a 60-watt light bulb, which is about 400,000 angstroms in diameter.

The most common approach to measuring surface area relies on the adsorption of gases on the coal surface. The amount of gas adsorbed at various pressures and the area of one gas molecule are used to calculate the coal surface area. The gases generally used in coal research are nitrogen and carbon dioxide. Unfortunately, tests of the same sample using the two gases show drastically different results. Surface areas calculated from carbon dioxide adsorption can be 10 times higher than areas calculated from nitrogen adsorption. The reason for this discrepancy is an issue of controversy. A possible explanation is that the very low temperatures used in experiments with nitrogen (320 degrees Fahrenheit below zero) slow the motion of the gas molecules enough to make it hard for them to diffuse into the pores. This explanation especially applies to so-called "ink bottle" pores, which have a very constricted opening but larger diameters in the interior.

An alternative technique is the measurement of the heat of wetting. The atoms or molecules on the surface of a solid are in a slightly higher energy state than the molecules in the interior. When the solid is wetted by a liquid, the molecules that were formerly the surface of the solid are not the surface any more. Consequently, they can "relax" to a lower energy state. The energy emitted by this relaxation appears as heat, that is, the *heat of wetting*. When a coal sample is added to a quantity of wetting liquid, the temperature of the liquid plus coal mixture rises. The temperature rise per unit area is determined from experiments with a standard substance that has been carefully characterized by an independent technique. In coal research, methanol usually serves as the wetting liquid, and an activated carbon sample serves as the standard.

Heat of wetting measurements are controversial because the total measured heat may include heat effects from the interaction of the methanol with oxygen groups on the coal surface. Rather than merely wetting the surface, some methanol may also be imbibed into the coal structure. Nevertheless, heat of wetting data are generally close to results for carbon dioxide adsorption. Lignites, which have abundant oxygen groups, give almost the same results in methanol as in the unreactive liquid cyclohexane.

Although the reliability of surface area data from a single technique is questionable, coal scientists generally agree that low-rank coals have the highest surface area. Surface area decreases as rank increases from brown coals and lignites to bituminous coals; then it increases again in anthracite. A lignite might have a surface area of about 150 square meters per gram; 90% of this surface area may be from the internal pore system. If all this surface area could somehow be unfolded and spread out, an ounce of lignite could cover a football field.

Density

Several types of density measurements are made, depending on the intended use of the data. The *bulk density* is the weight of lump coal held in a given volume. Knowledge of the bulk density helps engineers calculate, for example, the size of a storage bin needed to contain a given weight of coal or the weight of coal that could be held in a reaction vessel of a certain size. Bulk density is dependent on particle size. Many small particles of coal can be packed into a volume that would hold only a few large, irregularly shaped lumps. A rough, rule-of-thumb estimate is that the bulk density of coal is about 50 pounds per cubic foot. Packed in bulk, a given volume of coal weighs approximately 80% of what the same volume of water would weigh.

The measurement most often used to determine coal density is the *apparent density*. Measurement of apparent density is done by liquid displacement; the volume of liquid displaced by a known weight of coal must be equal to the volume of the coal. The apparent density depends on the liquid used to make the measurements because each liquid will penetrate the pores of the coal to different extents. Apparent densities range from about 1.4 grams per cubic centimeter (87 pounds per cubic foot) for coals of 80% carbon to about 1.3 grams per cubic centimeter for coals of 90% carbon. Coals of 95% carbon have apparent densities of 1.6 grams per cubic centimeter. (The density of water is 1.0 gram per cubic centimeter.)

Hardness

The *hardness* of coal is customarily measured by pressing a ball or cone onto the surface with a known, standardized force. This test leaves a dent on the surface; measuring the diameter of the dent gives an empirical measure of the hardness of the surface. On the Mohs scale of hardness, lignites range from 1 to 3, bituminous coals range from 2.5 to 3, and anthracites range from 3 to 4. For comparison, talc (as in talcum powder) has a Mohs hardness of 1, a penny has a hardness of about 3, and a pocketknife blade has a hardness of about 6. Because coal is heterogeneous and is filled with pores and cracks of various sizes, relating mechanical measurements of coal to fundamental properties is questionable. For example, we might suppose that hardness has something to do with the way coal breaks up during crushing or grinding. However, fracturing behavior seems to depend more on the presence of cracks or large pores. Hardness becomes an important consideration when pulverized coal is handled in suspensions, such as in coal–water slurry fuels or slurries of coal used in liquefaction. In these applications hardness is a concern because the coal particles can wear away internal parts of valves or pumps and lead to serious maintenance problems.

Grindability

The test commonly used to assess how easily coal can be ground involves grinding it with eight steel balls under a standardized load and number of revolutions. The weight of the coal that has been ground fine enough to pass through a 200-mesh sieve (which has openings three-thousandths of an inch in diameter) is used to calculate the *grindability*. The grindability is called the *Hardgrove index* in honor of the inventor of the test, Ralph Hardgrove of the Babcock and Wilcox Corporation. Bituminous coals have the highest Hardgrove index, meaning that they are the most grindable. Both anthracites and lignites are more difficult to grind. The Hardgrove index, even when run on different samples of the same coal, is influenced by the moisture content of the sample and the temperature. Despite the uncertainties built in by temperature or moisture variability, the Hardgrove test serves as a useful guide for calculating the sizes of grinding equipment needed to process coal.

Variability of Properties with Rank

Coal is a heterogeneous substance composed of lithotypes, macerals, minerals and other inorganic components, pores, and moisture. Despite decades of research by outstanding scientists, the fundamental nature of coal is still poorly understood. The structure of coal is still being studied and debated. The two common methods of measuring surface area by gas adsorption give results that differ by an order of magnitude. Coal is highly variable in composition and properties—lignite is certainly nothing like anthracite. Nevertheless, the composition and properties of coal often show regular patterns of variation with rank. As progress continues in basic research on coal, these patterns of variation will be better explained.

Chemical Composition

The ultimate analysis tells us that carbon content increases as rank increases. Hydrogen content is roughly constant as rank increases, until the low-volatile bituminous coals are reached; then hydrogen content decreases through the anthracites. Oxygen content decreases steadily as rank increases. Methoxyl ethers vanish at about 75% carbon. Methoxyl ethers are an important structural feature of lignin. As lignin becomes increasingly degraded in coalification, the methoxyl ethers are lost. Carboxylic acids occur only up to about 80% carbon. This fact has some important consequences for the composition of the ash. Phenol and carbonyl groups persist to 90% carbon; some bridging ethers may survive beyond this carbon content. Total sulfur generally increases as rank increases up through bituminous coals. Organic sulfur content is generally independent of rank; it is usually 1.0–1.5%. Sulfate sulfur is unimportant in most coals. Increases in pyritic sulfur are responsible for increases in total sulfur. Nitrogen content is also usually independent of rank and typically amounts to about 1%.

Aromaticity

Aromaticity of coal increases as rank increases and runs from about 0.6 in lignites to about 0.95 in anthracites. Also, the coalescence of aromatic rings increases slowly from 1 or 2 rings in lignites to 3–5 in bituminous coals and rapidly rises to more than 10 in anthracites. The attachment of groups on the aromatic ring systems decreases as rank increases. Because the oxygen groups that might attach to rings are also decreasing, fewer entities occur in the coal to attach to the ring systems. As the number of rings fused together grows, more of the carbon atoms become buried inside the rings and fewer are available as sites for attachment. For example, the two-ring system of naphthalene contains 10 carbon atoms of which 8 could, in principle, be attached to other groups. In other words, 80% of the carbon atoms are available. In the four-ring structure of chrysene, 67% of the carbon atoms are available for attachment to other groups (12 of 18); in the four-ring structure of pyrene, only 62% are available (10 of 16). A hypothetical 16-ring structure in anthracite would have a total of 48 carbon atoms, but only 18 would be potential sites for substitution (38%). As these very large aromatic ring systems begin growing in the transition from bituminous coal to anthracite, more and more carbon becomes incorporated into the interior, leaving fewer places for hydrogen. Thus, anthracites should contain less hydrogen than bituminous coals, and this expectation is confirmed by ultimate analysis data. The removal of oxygen groups and increased aromaticity as coalification proceeds are consistent with the growth of aromatic ring systems, the decrease in substitution on the rings, and the decrease in hydrogen content.

Moisture Content

The proximate analysis shows a general tendency for moisture content to decrease as rank increases; however, a slight increase in moisture content occurs again in anthracite. This observation ties in with two other factors: First, the oxygen groups are decreasing as rank increases. This loss of oxygen removes sites from the coal where water could be chemically attracted through hydrogen bonding.[7] Second, the porosity decreases as rank increases from brown coals and lignites to bituminous coals, resulting in fewer holes that could be filled with water. The increased porosity of anthracites compared with bituminous coals provides some additional spaces for water, and indeed a slight increase in moisture content is observed.

Volatile Matter

Volatile matter decreases as rank increases. As the rank of coal increases, the structure changes as methoxyl ethers, carboxylic acids, and aliphatic carbon atoms are removed. These structural features are the most likely to break apart during heating and be driven off as volatile matter. As they are removed, the big aromatic ring systems increase in size. The aromatic ring structures are much less likely to break apart or vaporize; in a sense, they are fixed in the coal. Thus, fixed carbon increases as rank increases. (Although this increase in fixed carbon can be explained in terms of chemical structures, it also reflects a mathematical artifact: By definition, on a moisture- and ash-free basis, if volatile matter decreases, fixed carbon must increase because together they must sum to 100%.)

Ash Content

The *ash content* of coal is generally quite variable and does not reflect the rank of the coal. Variations result from differences through geological history and from place to place in the amounts of detrital mineral matter carried into the coal. However, the *ash composition* does show some rule-of-thumb relationships to rank. Higher than about 80% carbon (i.e., bituminous coals), the carboxylic acid groups no longer contribute to the coal structure. However, these were the groups that provided the sites for incorporating much of the total amounts of sodium, calcium, and magnesium in the coal. Consequently, the ash compositions of the bituminous coals tend to have proportionately more silicon, aluminum, and iron and less sodium, calcium, and magnesium than the low-rank coals. Other than these rule-of-thumb relationships, the ash composition does not vary in a regular way with rank for the same reason that the ash content did not: The potential for adding inorganic constituents to the coal varied greatly over the course of time and in ways that depended on the local geology.

Heating Value

The heating value of coal increases up to the low-volatile bituminous coals. This trend relates to the loss of oxygen groups. A carbon atom that is bonded to an oxygen atom is already partially oxidized. Less heat will be gained by converting that carbon atom to carbon dioxide during combustion than by converting a carbon atom that is bonded only to other carbon atoms or to hydrogen. Although anthracites are the highest ranking coals, they do not have the highest heating values. In anthracites the aromatic ring systems have gotten so large that some of the energy that might have been released in combustion is actually used to break down the aromatic bonding in the ring systems.

Conclusion

Many of the physical properties of coal show a consistent variation with rank. If density, surface area, porosity, or grindability are plotted as a function of the carbon content, all of these properties either hit a minimum or a maximum in the region of 85–90% carbon and then change direction. For example, surface area drops steadily from lignites, hits a minimum in the bituminous coals around 87% carbon, and then rises again in the anthracites at about 95% carbon. The same pattern of variability of seemingly unrelated physical properties strongly suggests an explanation relating to the three-dimensional molecular structural arrangements. As the aromatic ring size increases and the ring substitution and number of comparatively bulky oxygen groups decrease, the structural units should more easily move closer together. The large aromatic systems in anthracites are so big and inflexible that they are difficult to fit into a tightly arrayed structure; thus, in a sense, the structure "opens up" again. As the carbon content of coal increases to 85–87%, the number of cross-links in the macromolecular structure decreases. Because the oxygen content of coal decreases as carbon content increases, bridging ether groups may possibly participate extensively in the cross-linking in coals of less than 85% carbon. The progressive decrease of cross-linking reflects a loss of ether oxygen atoms. As carbon content increases beyond 88%, the amount of cross-linking increases rapidly. The cross-linking in coals of high carbon content is not due to ethers to a great extent. The hydrogen content begins to drop as carbon content exceeds 90%. Thus, the cross-linking in coals containing more than 88–90% carbon may arise from the loss of hydrogen atoms as more carbon-to-carbon bonds form.

Characterizing the heterogeneous and variable substance called coal and constructing a framework of understanding is a difficult task. J. E. Gordon, in his recent book *The New Science of Strong Materials*[8], comments on the discipline of materials science:

> *I think it is fair to say that we now understand a great deal more about the reasons for the mechanical behavior of solids than we did only a very few years ago. This may be because many of the raw materials of understanding were lying to hand already. There was a great body of orthodox physical and chemical knowledge and there was also, although in different hands, an accumulated mass of engineering experience and tradition. To fit them together and to make one explain and confirm the other required only a moderate amount of original experiment and fresh thinking.*

Coal science is not yet as mature a discipline as materials science. Yet, coal researchers have an analogous stock of "raw materials of understanding", that is, physical and chemical background and accumulated engineering experience. These raw materials are being transformed to increased knowledge of the substance called coal.

Notes

[1] Stein, G. *The Autobiography of Alice B. Toklas*; Vintage Books: New York, 1961; p 165.

[2] Several nomenclature systems are used in coal petrology. The terminology used here follows the recommendations of the International Committee for Coal Petrology.

[3] The extraction of waxes from lignite is a commercial process. The products are called *montan waxes* and are used as components of shoe polish, furniture polish, and waterproof paints.

[4] van Krevelen, D. W. Plenary lecture presented at the 8th International Conference on Coal Science, Dusseldorf, Germany, September 1981.

[5] This statement assumes that the structural formulas are consistent with the ultimate analysis and other characterization data.

[6] Because density is the ratio of mass to volume, measuring the density of a sample of known mass provides a straightforward way to calculate volume.

[7] A hydrogen bond can occur when a hydrogen atom bonded to an atom of a very electronegative element—usually fluorine, oxygen, or nitrogen—experiences an electrical attraction to a second fluorine, oxygen, or nitrogen atom. The hydrogen bond is much weaker than the normal chemical bond between two atoms.

[8] Gordon, J. E. *The New Science of Strong Materials*; Princeton University: Princeton, 1976; p 258.

Chapter 5

Coal from the Ground

Mining Preparation

Locating a Deposit

Before coal mining can begin, a deposit of sufficient size and quality for profitable mining must be located. Finding coal begins with a thorough geological survey of the region in which coal is suspected or hoped to be found. An examination of fossils in the region helps to determine the geologic age of the local formations. Most important commercial coal deposits in the United States are either of Carboniferous or late Cretaceous–early Tertiary age. Formations not of these ages are much less likely to contain coal strata. *Faults* (fractures in the rock in which one side has been displaced relative to the other) may move a coal bed either vertically or horizontally. The existence of faults may make it very hard to determine the continuity of the coal seam. Usually coal occurs in formations that have not been severely altered by metamorphism. Therefore, regions of igneous or metamorphic rock are unlikely to contain coal. One very important feature in coal exploration is the existence of coal at or very near the surface. Such outcrops of coal can help the geologist estimate the thickness of the seam and its *pitch* (the angle at which the seam inclines below the horizontal).

Drilling Bore Holes

When the existence of a coal deposit seems likely, either by observation of coal outcrops or by inference from other geological information, the next step is to drill a series of bore holes. The tool used for this operation is a

1171–X/87/0093$07.00/0

core drill, which is a hollow bit made of steel pipe. The bit has cutting teeth around its lower edge and is attached to a core barrel. One type of core drill is the diamond drill, on which diamonds are mounted in the cutting end, set to protrude a short distance beyond the inner and outer edges of the pipe. (The diamond drill is constructed this way for the same reason that the teeth of a saw blade are staggered slightly: The saw will cut a kerf just a bit wider than the blade itself.) The drill makes a ring-shaped cut through the rock, forming a core of rock that is retained in the barrel and is removed from the hole with the bit. The drill cores provide much useful information for assessing the nature and extent of the deposit. The coal samples brought to the surface in the cores are analyzed to determine the quality of the coal, particularly those coal properties that would affect its use: moisture, sulfur, ash, and heating value. A series of holes drilled throughout the region provides data that can be assembled into a picture of the total extent of the deposit, seam thickness, pitch, and other types of rock that will have to be drilled to reach the coal.

Electric Logging

Besides studying drill cores, geologists can also study *electric logs*. In this procedure, instruments are lowered down the hole to send back a series of electrical impulses that help describe the various rock layers. An *electric log* is a chart of two electrical properties—spontaneous potential and resistivity—as a function of depth in the hole. An electric potential exists between rock exposed in the bore holes and the fluid used to lubricate the drill. The magnitude of this potential is affected by how well the fluid can penetrate the pores of the rock. Thus, the spontaneous potential is a measure of the permeability or porosity of the rock. Virtually all sedimentary rock formations contain some water, which is made somewhat conductive by the presence of dissolved salts. Conductivity, and therefore resistivity, is affected by the composition of the rock, its porosity, and the concentration of salts dissolved in water held in the rock pores. By comparing the spontaneous potential and resistivity charts with the behavior of known rocks, the geologist can identify the kinds of rocks through which the hole passes. Aerial photographs are another tool for coal exploration; they are used to help prepare surface maps of the region. Coal exploration today is heavily reliant on sophisticated electronic and analytical instrumentation for obtaining detailed information on the regional geology and coal quality. It is a far cry from the rough-and-ready days of coal prospecting described in Frank Kneeland's 1926 book, *Preliminaries of Coal Mining*[1]:

> *The man who undertakes work of this kind should have a reasonably full knowledge of the geology of coal and the formation in which it may be found.... He should know where Carboniferous deposits are likely to be found and, what is*

> *possibly of more importance, he should also know where they are not likely to be found. He should possess sufficient knowledge of the use of tools to enable him to build such simple structures as his work may require and should be enough of a blacksmith to be able to sharpen his picks and chisels and, if necessary, shoe a horse.*
>
> *...When operating in a settled country he will require only the tools and instruments necessary to the prosecution of his prospecting work proper. If the country in which he operates is settled sparsely, he will require additional impediments, including even weapons. In its simplest form his outfit should consist of a compass, a clinometer, a pick, a shovel, and, in most instances, a set of drills, a hammer, a spoon or sludge pump for cleaning holes, a quantity of explosive, suitable means for detonating explosive charges, an aneroid barometer, and a prospector's pick or hammer.*
>
> *If the region is well-watered and traversed by numerous streams, a canoe will be found handy. If no streams, lakes, or marshes of any extent are present, a pack pony or mule may be substituted, while in some cases packers, who will carry the outfit upon their backs...must be relied upon. The weight that a man will carry in this manner will range from 60 to 120 pounds.*
>
> *The outfit should consist of a light tent...cooking utensils, a supply of food sufficient to last, say, 1 month, one ax or heavy hatchet, and four blankets. It would also be well for the prospector to provide himself with two compasses.... He should also have two knives—a sheath knife and a heavy, businesslike pocketknife with from two to four blades.*

The drilling phase of a coal exploration program can be very expensive. Depending on the extent of a deposit and how much it appears to vary throughout the region, a mining company might drill hundreds or even thousands of holes as part of its exploration program. The actual drilling is often a very slow process; progress sometimes is made at rates of only 1–2 feet per hour. A coal seam 1000 feet below the surface could take 3–6 weeks of round-the-clock drilling to be pierced by one bore hole.

Seismic Surveying

An exploration technique of long standing in the petroleum industry now starting to be used in coal exploration is *seismic surveying*. A series of holes about 10 feet deep is drilled, loaded with an explosive, and fired sequentially. The series of explosions generates a set of shock waves that are reflected back to the surface from the layers of rock as they travel through the ground. The reflected waves are picked up by detectors called *geophones*.

Different types of rock reflect the shock waves with different intensities. The data assembled from the geophones give information about the types of rock layers and any subsurface structures such as fractures or folds in the rock. This information can be useful in determining shifts in direction of the coal seam.

Interpreting Data

Once the data from the exploration program have been gathered, some difficult questions must be answered. The answers can result in investing millions of dollars of corporate resources on the potential for successful mining.

- How much coal is there?
- Do the thickness and nature of intervening rock layers make it prohibitively difficult to reach the coal?
- Are the coal seams thick and wide enough to be mined by available mining technology?
- Is the quality of the coal such that it can be marketed or fairly easily upgraded to market specifications?
- Is the coal likely to be worked for enough years to repay the costs of opening the mine and to yield a profit for the company?

If these questions can be answered, a decision can be made whether or not to proceed with mining operations.

Underground Mining

Coal mines fall into two broad categories: underground mines and surface mines. In most instances a selection between underground mining and surface mining (most often strip mining) is based on the depth of the coal seam to be worked. Seams lying near the surface are accessible by stripping, but this method could be prohibitively expensive and disruptive to the surroundings if used for mining deep-lying seams. Each mining method has its own set of technical and economic advantages and disadvantages. Each method also has detrimental effects on the miners and the environment.

Classification

Underground mines are classified by the following types of access used to reach the coal:

- shaft mines,
- slope mines, and
- drift mines.

These classes are shown in the sketch in Figure 5.1. The type of access is not determined by the extent of the coal deposit or by the production of the mine, but rather by the depth and pitch of the seam. In practice, large mines might use two or even all three of these methods. Mines are required to have at least two openings to aid ventilation and to provide a second exit for miners in case of trouble.

Shaft Mines. A *shaft mine* uses a vertical hole dug straight down from the surface to the coal seam. Tunnels, or drifts, are then driven horizontally into the coal. Miners and equipment are lowered in a cage by a hoist to the working level. If more than one seam is being worked, the drifts may exist on several levels. The coal is loaded into tram cars that run along railroad tracks to the shaft, and then it is hoisted to the surface. A sump can be dug at the bottom of the shaft to collect water, which is then hoisted or pumped to the surface. Shaft mines provide the most straightforward access to deep-lying seams but have the disadvantages of the expense and upkeep of the hoisting machinery.[2]

Slope Mines. A *slope mine* provides an access at a slant. A slope may be driven to follow a seam along its pitch or to cut through a hill or mountainside to reach coal underneath. The coal can be brought to the

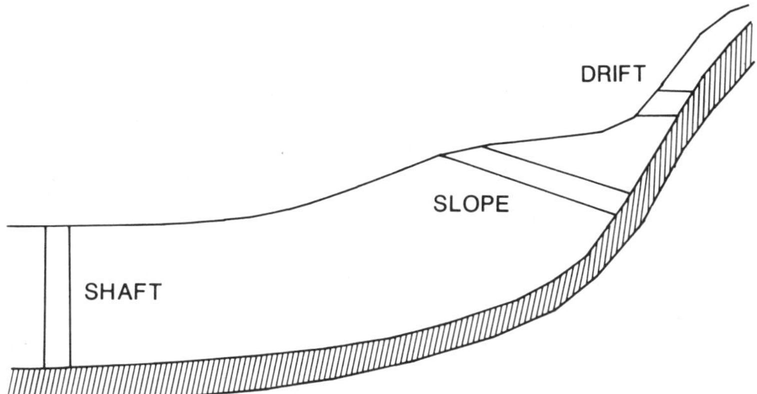

Figure 5.1. The type of access to a coal seam depends partly on the local topography and how close the seam lies to the surface. A deep-lying seam can be approached via a shaft mine. A slope mine can be driven into a hillside. Steeply pitching coal seams can be reached by a drift mine cut straight into the hill.

surface on conveyors. Compared with shaft mines, slope mines have the advantages of easier removal of the coal and easier movement of miners and machinery into the mine.

Drift Mines. Where a coal seam outcrops on a mountainside, a *drift mine* can be driven straight into the coal. Of the three types of underground mines, drift mines are by far the easiest and cheapest to develop. The coal can be brought out on conveyors or on tram cars pulled by small electric locomotives (or, in the early days, pulled by mules). Drift mines are used where the coal slopes upward from the outcrop. Driving upward into the coal allows the uphill grade to provide a natural drainage for mine water.

Coal Cutting Systems

Coal is cut from the seam by one of three systems:

- room and pillar mining,
- pitch mining, or
- longwall mining.

The type of system used does not depend on the type of mine access.

Room and Pillar Mining. The *room and pillar system* is used when seams lie relatively level. In the room and pillar system, the coal is divided into a series of blocks. Parallel tunnels called *main entries* are driven into the coal. The main entries are connected by a second series of tunnels, called *gangways*, which are cut at right angles to the main entries. Then, working from the gangways, the miners cut the coal into rooms or chambers at intervals along the gangway. A portion of the coal is taken out as the rooms are formed. The rest of the coal is left in place as solid pillars to support the rocks above the mine working (the roof). Coal production in the room and pillar system depends on the height and width of the face of the seam being worked. The height is determined by the thickness of the seam. The width is the maximum distance that can be cut away without the roof collapsing.

Driving the main entries and gangways is called the *first working*. If mining stops when the extent of the seam is reached, about half the coal will have been recovered. The actual percentage of coal recovered depends on the number and size of pillars needed to support the roof. The need for roof support is determined by (1) the safety of the miners and (2) the prevention of *subsidence* (falling, lowering, or flattening out) of the surface, which could be caused by collapse of the roof. Whether or not subsidence is an important consideration is determined by the surface activity. If the mine is being worked under a highly populated area, subsidence must be prevented; however, in unpopulated regions it may be of little consequence.

Stopping operation after the first working leaves considerable coal in the mine. Room and pillar mining can be very wasteful unless the pillars are removed. Whether or not the pillars will be taken out is determined both by economic factors and by the need to prevent subsidence. The removal of the pillars is technically known as the *second working* but is also referred to as the "retreat" or as "robbing the pillars".

The progressive removal of roof supports as the pillars are robbed makes the second working very hazardous to the miners. Robbing the pillars can weaken the roof enough to cause collapse. The necessity of leaving pillars for roof support was tragically demonstrated by a disastrous cave-in at the Twin Shaft Colliery in Pittston, Pennsylvania, in 1896. Fifty-eight men died while trying to shore up an unsupported roof that had been showing signs of cracking. The commission that investigated this accident strongly recommended leaving the pillars in place as the coal is mined.

The actual mining of the coal occurs by one of two processes: conventional mining or continuous mining. About one-third of the coal produced in underground mines is obtained by conventional mining. Conventional mining replaced digging coal by hand in the 1930s and was the predominant mining method into the 1950s. It is slowly being supplanted by continuous mining.

Conventional Mining. In conventional mining, a machine cuts a slit a few inches wide and several feet deep into the base of a coal seam. A second machine then drills a series of holes into the face. The holes are packed with explosives. When the explosives are fired, the blast shatters the coal and the broken pieces fall to the floor because of the slit undercutting the seam. (An alternative to explosives is cartridges containing air or carbon dioxide at about 1400 atmospheres of pressure; the sudden release of gas in the hole shatters the coal.) Miners load the coal onto tram cars or conveyors for haulage out of the mine. The miners arrange support for the newly exposed roof, and the cycle is begun again. A work crew of 8–15 miners can mine 1000 tons of coal in a single shift.

Conventional mining is particularly advantageous in situations in which the seam thickness or pitch changes frequently. It would be difficult to use the large machinery required for continuous mining in such situations. Conventional mining also produces lesser amounts of fine-sized coal and coal dust than other methods. Dust is particularly undesirable because it is a waste of coal, a health hazard to the miners who breathe it, and a potential explosion hazard. On the other hand, the cyclic nature of conventional mining means that coal is only being extracted part of the time. The rest of the miners' time is spent cutting, drilling, and propping the roof. Conventional mining becomes particularly inefficient when the roof is weak for some distance because then considerable time and effort must be devoted to supporting the roof. Although conventional mining is not as labor-intensive as mining by hand, it does use more miners than the more highly mechanized continuous mining. On the basis of coal production per miner, conventional mining suffers in comparison with continuous mining.

Continuous Mining. In continuous mining, a single machine that creeps along on caterpillar tracks uses cutters to rip or gouge coal from the face of the seam. The coal is automatically loaded onto conveyor belts or into cars. One miner operating a continuous mining machine can produce up to 12 tons of coal per minute. Continuous mining now accounts for about two-thirds of U.S. underground coal production.

Regardless of the type of mining, the roof must be supported, adequate ventilation must be provided, and drainage systems for mine water must be installed. The continuous miner can cut coal so rapidly that the workers attending to roof support, ventilation, and drainage can't keep pace. The mining machine must be stopped from time to time to allow the rest of the work to catch up. The introduction of continuous mining machinery in the late 1940s also led to the introduction of readily movable and extensible conveyor belts because other methods of moving the coal cut away from the face were too inefficient to use with continuous mining.

The advantages of continuous mining compared with conventional mining are the reduced labor costs and the complete elimination of expenses for explosives and cutting and drilling machinery. However, continuous mining is not effective when the coal is extremely hard. Also, the machinery may be too unwieldy to use when the pitch and thickness of the seam vary.

Pitch Mining. *Method.* Pitch mining is a technique used when the coal seams are inclined. Pitch mining is frequently used in anthracite mining; in extreme cases, anthracite seams might be sloped at 80 degrees to the horizontal. A gangway is driven along the bottom of the seam. At intervals, cuts are made at right angles to the gangway, passing up into the seam. Most of the space created by the cuts is used to build a timbered chute. Space is left on either side of the chute for a passageway. The bottom of each chute, opening into the gangway, is closed with a heavy, timbered bulkhead.

The miners climb or crawl up the passageway to reach the face of the seam. They must carry or drag with them their drills, explosives, roof supports, and some lumber to extend the chute. The coal is drilled and blasted, and the coal loosened by the explosion is shoveled into the chute. The newly exposed roof is supported, and the sides of the chute are extended. Mining can proceed at a rate of about 6 feet per day into the coal seam. In the case of steeply sloping seams, the most strenuous part of a miner's shift can be getting from the gangway up to the face of the seam with a complete load of equipment and supplies.

The bulkhead at the bottom of the chute is equipped with a gate. Opening the gate allows loose coal to fall by gravity into tram cars in the gangway to be hauled out of the mine.

Blasting and Explosives. Blasting is the principal method of fracturing and loosening the coal in both conventional and pitch mining. Explosives were not used in the early days of coal mining in the United States. Instead, a hole was drilled into the coal and was packed with quicklime (calcium oxide). When the quicklime absorbed water, it expanded and fractured the coal. Wedges were then driven into the cracks to split off pieces of coal.

Black powder, made from charcoal, sulfur, and potassium nitrate, was introduced to mining in 1818 in Plymouth, Pennsylvania. During the 19th century, black powder displaced other methods and became the predominant explosive in coal mining. To blast or shoot off coal, miners first drilled a hole into the seam and then pushed a powder cartridge into the rear of the hole with a tamping rod. Then they pushed a long, thin copper rod—the *miner's needle*—into the hole and packed the hole with clay or wet coal dust. When the miner's needle was carefully removed, a thin hole was left running back to the powder. Now, a slow fuse or a squib[3] was inserted. The fuse or squib was lighted, everyone took cover, and, if all went well, the resulting blast shattered the coal.

This system worked most of the time. However, many miners were seriously injured or killed in blasts that occurred before the miners could reach shelter. An explosion too powerful could cause roof rock to fall. The flame from the explosion of the blasting powder could ignite coal dust or flammable gases in the mine and thereby cause fires or second explosions far more devastating than the blasting of the coal.

Dynamite was invented in 1867 by Alfred Nobel and is made by absorbing nitroglycerin into sawdust, wood pulp, or a chalky material called *kieselguhr*. It began to be used during the 1870s. Dynamite was fired with fuses or squibs until about 1920, when electric firing began to be used.

The danger of the flame of the original explosion setting off a fire or a secondary dust or gas explosion led to the development of *permissible dynamite*. This material contains a high proportion of ammonium nitrate and relatively little nitroglycerin. It produces only a brief, short flame and is safer than ordinary dynamite or black powder. Permissible dynamite was introduced in 1909, and its use in underground mining is now regulated by the Bureau of Mines.

Timbering and Roof Bolting. The roof is supported by timbering or roof bolting. *Timbering* is the use of wood to support the faces of an excavation. The timbers do not support the full weight of all the rock overhead; this weight is supported by the rock or coal forming the sides of the tunnels. The purpose of timbering is mostly to keep relatively small, loose pieces of the roof from falling. Although the history of underground coal mining sometimes appears to be a sequence of tragedies killing dozens or hundreds of miners, far more deaths have occurred as a result of small falls of rock killing one or a few miners at a time.

The form of timbering depends on the condition of the roof. The simplest method is a single vertical prop with a cap piece. Where the sides as well as the roof may be weak, two vertical props supporting a horizontal collar between them are used. Such sets of timber might be erected every few feet. Soft woods are often preferred as props because they are relatively strong but elastic enough to bend somewhat before breaking. The bending of the props warns miners of an impending *squeeze* (a condition in which the weight of the roof rock suddenly crushes the coal as the roof settles down to the floor); a squeeze could be catastrophic to miners caught nearby. Today, timbers are sometimes made of masonry, concrete, or iron.

When support of the roof is no longer needed, the timbers are sometimes drawn, that is, removed. In principle, this removal allows the roof to settle gradually and regularly and relieves pressure on the coal or rock in the sides of the tunnel. This practice can allow surface subsidence to occur, which is particularly undesirable in populated areas. Furthermore, drawing the timber can be an extremely dangerous operation for the miners because the removal of one prop might initiate a collapse of the roof.

Roof bolting consists of drilling holes into the roof, inserting steel rods—the roof bolts—into the holes, and anchoring the bolts in place. Several thin, weak layers of rock are bolted together. Roof bolting helps to hold the rock in place. It also strengthens the roof because the thin rock layers of various strengths are effectively bolted together to make one thick compound beam that is stronger than any of its individual components.

Longwall Mining. *Longwall mining* is the removal of coal from one long, continuous face rather than removal from a number of short faces as occurs in room and pillar or pitch mining. Longwall mining is practiced in two versions: advance and retreat. In either case, mining is begun by driving two parallel roadways into the seam. These roadways can be up to 650 feet apart. The roadways are connected by a tunnel driven at right angles to them; this tunnel forms the face of the seam and is where the mining operation takes place. In *advance longwall mining*, the tunnel is driven at the front of the seam, and mining operations extend progressively deeper into the coal. In this case, the parallel roadways are continuously extended as well. In *retreat longwall mining*, the roadways are driven to the full extent of the seam and then connected with a tunnel. Mining moves from the back of the seam toward the main shafts.

Removal of the coal is done with an electrically powered machine that moves back and forth across the whole extent of the coal face. The machine can be equipped with heavy revolving chains carrying tungsten carbide tipped cutter picks or with a hardened blade somewhat like that of a plow. The mining machine can cut from a few inches up to 1.5 feet into the coal on each pass at a rate of up to 5 tons of coal per minute. The cut coal is directed onto a conveyor belt for transportation out of the mine.

The roof directly over the mining operation is supported by movable hydraulic pillars. As mining advances (or retreats), the pillars are moved so that the roof is always supported in the work area. The previously exposed roof is no longer supported and is allowed to collapse. This system provides some advantages for longwall mining: It eliminates some of the cost of permanently installed roof supports, and it allows working at depths at which the pillars left by room and pillar mining do not support all the weight of the overhead rock. Longwall mining is probably a safer system in mines in which the strength of the roof is generally poor.

Longwall mining requires long, reasonably level blocks of coal. It has been used on coal seams ranging from 1 foot to about 50 feet in thickness. It has also been adapted for sloping seams, although it never became very popular in the anthracite region, where seams can have extreme slopes. When the block of coal being worked is nearly depleted, a new area must be prepared for mining. Moving the machinery to the new area can cause a costly and time-consuming delay in the operations.

Although longwall mining is gradually replacing the other methods used in the United States, it still accounts for a minority share of the total underground coal production. A variation called *shortwall mining* has been introduced. The basic idea is the same as in longwall mining, but the working face is shorter, on the order of 150–200 feet, and the coal is cut with a continuous-mining machine. Longwall mining has become the predominant technique in many countries. Longwall mining accounts for about 75% of underground production in the Soviet Union and almost 90% in Great Britain. The advantages are (1) versatility in application to various seam thicknesses and pitches and (2) a high percentage of extraction of the coal; these must be balanced against a high investment cost for the mining equipment and occasional needs for costly moves to new locations.

Problems

Regardless of the mining system used, three problems must be faced in every underground mine:

- dust,
- gases, and
- water.

Dust. In regard to the mining operation itself (that is, disregarding potential long-term health consequences), the problem caused by coal dust is the potential of its fueling a devastating explosion. When fine dust particles of any flammable material are suspended in air, an explosion can occur. Grain elevators and flour mills, for example, have been the scenes of serious explosions. All else being equal, the rate of reaction between a gas and a solid will increase as the size of the solid particles decreases. The rate of reaction will also increase as temperature increases. If a suspension of combustible coal dust in air is ignited by even a small spark, a rapid combustion that raises the temperature occurs. This increase in temperature increases the reaction rate, which in turn raises the temperature even more. The whole dust suspension thus burns violently in a fraction of a second, an event that is observed as an explosion. Coal dust explosions have probably occurred everywhere that coal is mined. The worst happened in Manchuria in 1942 and claimed 1572 lives.[4]

Methane was long presumed to be the cause of devastating explosions; it is known colloquially to miners as *firedamp*. Mixtures of methane and air can explode if the concentration of methane is sufficient. The chain of events leading to such explosions was deduced in 1845 by Michael Faraday, who started his career as an apprentice bookbinder and was eventually recognized as one of the greatest scientists of the 19th century. Faraday observed that a first methane explosion, which might be relatively localized in a mine, serves to raise a cloud of coal dust and ignite it, producing a second explosion that is more widespread and far more violent. In addition, one of the products of the combustion reaction producing the dust explosion is carbon monoxide, a toxic gas that can be fatal to workers still in the mine or to rescuers entering the mine.

The concentration of coal dust that is suspended in air and can explode ranges from less than 0.1 ounce to 1 ounce of dust per cubic foot of air. In a tunnel 7 feet high and 8 feet wide, for example, about 0.25 pounds of coal dust per linear foot of the tunnel would be enough to explode. Even if the air is saturated with moisture, an explosion could still occur if the proper amount of dust is present.

Two approaches are taken to avoid dust explosions. The first is to attempt to minimize the production of dust during mining operations and to minimize sparks or open flames. This concern led to the introduction of permissible dynamite as a replacement for black powder or conventional dynamite.

The second approach is a technique called *dusting*. An incombustible dust, such as powdered limestone, is spread in the mine. In some cases, the mine is dusted just before blasting. The deliberate explosion of dynamite (or the accidental explosion of methane) raises a cloud of dust; however, the stone dust dilutes the coal dust cloud to the point at which the mixture in air is nonflammable. Some mines have used dust barriers to control the propagation of an explosion. A dust barrier consists of several tons of stone dust on a tilted platform. The shock wave of a coal dust explosion blows the massive quantity of stone dust into the air. The dense stone-dust cloud that results absorbs some of the energy of the explosion, reduces the temperature of the flame, and dilutes the coal dust.

Gases. Mine ventilation is required to provide the workers with fresh air and to remove mine gases that are potentially explosive or hazardous to health. The flow of air occurs through air shafts that are usually linked in pairs as an intake and a return. The shafts or drifts used to enter the mine sometimes serve as the intakes. The air moves from the intake across each working face and then out through the return. Federal mining law requires an air velocity of 3000 cubic feet per minute across the working face. Originally, natural convection currents, set up by temperature differences between the air in the mine and surface air, were the only source of ventilation. Another early approach was to install furnaces over the return; the heated air rising out of the furnace would draw air through the mine. Today, large fans are used to provide a positive ventilation system. Large mines can require fresh air at rates exceeding 0.5 million cubic feet per minute.

The gases of concern to miners are collectively known as *damps*; this term is a corruption of the German *dampf*, meaning vapor. The following five damps are hazardous to miners:

- firedamp,
- whitedamp,
- blackdamp,
- stinkdamp, and
- afterdamp.

Firedamp. Firedamp, or methane, is the most troublesome damp. It is a product of coalification and is frequently found absorbed in coal and the surrounding rocks. Normally, it is given off gradually as the coal or rock is removed in the mining operations. The amount of methane varies from one coal deposit to another. Sometimes as much as 5000 cubic feet of methane is liberated for every ton of coal dug.

Methane is odorless, tasteless, and colorless, so its presence is not obvious. Mixtures of 5–14% methane in air are explosive. Before the early 19th century, miners used candles or lamps with open flames for illumination. If the concentration of methane in the mine reached a dangerous level, an explosion could be set off by the miners' lamps. By the close of the 18th century, the incidence of mine explosions made it increasingly clear that some device was needed to provide safe illumination underground. Several inventors, among them George Stephenson, better known as the inventor of the railroad locomotive, developed safety lamps. The early designs were unwieldy or difficult to use. In 1815, the Society for Preventing Accidents in Coal Mines asked for the help of one of England's most eminent chemists, Humphry Davy.[5] The invention of a convenient and practical safety lamp by Davy in 1816 was a tremendous boon to miners and has saved an incalculable number of lives.

The important design feature of Davy's lamp was the metal gauze surrounding the flame. The gauze had a very high thermal conductivity and dissipated the heat of the flame enough to prevent ignition of gas outside the gauze. Although any gas that diffused through the gauze would burn inside the lamp, no danger of a gas explosion was present. The original Davy design was modified many times, but the principle remained the same. These safety lamps were eventually replaced by battery-powered electric lamps for illumination, but they can still be used as gas detectors.

Although methane is usually given off gradually, occasionally a large quantity is liberated almost instantaneously in phenomena called *outbursts*. Outbursts can be very dangerous because they may be accompanied by the violent ejection of broken rock fragments and large amounts of dust. A spectacular outburst occurred in the Morissey mine in British Columbia shortly after the beginning of the 20th century. About 2000 tons of coal was blasted loose by the sudden release of 3 million cubic feet of methane.

Whitedamp. Whitedamp is the miners' name for carbon monoxide. Carbon monoxide is a toxic gas; it reacts with hemoglobin to block its ability to transport oxygen through the bloodstream. About 0.5% of carbon monoxide in air can be fatal in 2 minutes. Whitedamp is also a combustible gas, thus potentially contributing to fires or explosions. Sources of carbon monoxide in mines are slow oxidation of the coal, incomplete combustion of explosives, and exhaust gases of diesel engines.

Blackdamp. Blackdamp, also called *chokedamp*, is chiefly carbon dioxide. Although it is not poisonous, it can cause asphyxiation by displacing breathable oxygen. Blackdamp is particularly dangerous because it is heavier than air and thus tends to accumulate in low spots of mines. Miners entering low-lying spots may not at first be aware of any danger until they are overcome by a lack of oxygen. Blackdamp is formed in combustion or oxidation processes: lamp flames, explosions (whether deliberate or accidental), coal fires, rotting of timbers, and respiration of the workers.

The safety lamp can serve as a convenient warning detector for firedamp and blackdamp. Firedamp is lighter than air; thus, it will accumulate in high spots. By raising the lamps toward the roof, miners can check for firedamp by looking for a characteristic lengthened flame pattern of the lamp when firedamp is present. Blackdamp, on the other hand, is heavier than air and is incombustible. If the flame goes out when the lamp is lowered toward the floor, then blackdamp is present.

Stinkdamp. Stinkdamp is hydrogen sulfide, a gas that is infamous as the unforgettable odor of rotten eggs. Although the odor is the most obvious characteristic of hydrogen sulfide, it is also a powerful poison and combustible gas. The odor is so strong that hydrogen sulfide can be detected at concentrations as low as two-millionths of an ounce per cubic foot of air. Hydrogen sulfide fatigues the olfactory nerves so that they stop responding to its odor; thus, the nerves then fail to give a warning of higher concentrations. A concentration of 0.1% in air can be fatal in a few moments. Stinkdamp is formed when acidic mine water reacts with pyrite in the coal or when waste left in worked-out areas of the mine catches fire. Some hydrogen sulfide also occurs with methane as a result of natural processes.

Afterdamp. Afterdamp is a mixture of gases remaining after an explosion or fire. The composition of afterdamp varies, depending on what burned or exploded and under what conditions. However, afterdamp is always very deficient in oxygen and contains considerable amounts of carbon monoxide. Miners who are not killed instantaneously in an explosion might be killed by the afterdamp before they are reached by rescuers. The rescuers themselves can be in danger from afterdamp when they enter the mine.

Water. Water can be a constant problem in some mines. Water can enter the mine by running in from the surface or from seepage from underground streams. To cope with the water, ditches can be dug along the sides of the gangways to drain the water into one collection area, the *sump.* The water can then be hoisted out in tanks or pumped out of the mine. In extreme cases, as much as 20 tons of water have to be removed for every ton of coal dug. Mine water can be highly acidic. The oxidation of pyrite in the coal can lead, through a series of reactions, to the formation of sulfuric acid. Acidic mine water is very corrosive, so pumps and piping systems must be made of special alloys or lined with acid-resistant materials such as lead or bronze. For many years the acidic mine water was discharged into nearby streams or rivers. This practice caused extensive damage to the environment because aquatic life could not survive in the acidic water. Current regulations require that the water be neutralized before it is discharged.

The need to keep mines dry caused a chain of events that had an incalculable effect on the course of industrialization. Problems of removing water from mines were already severe in the 17th century. In 1698, Thomas Savery, an English military engineer, invented a steam-powered pump called the *miner's friend.* An ordinary lift pump uses the pressure of the atmosphere to force water upward. The distance over which the pump can lift is limited to the height of a column of water supported by the atmosphere, or 33 feet. Savery overcame this limitation by designing a pump in which the water filled a closed vessel and was then expelled by high-pressure steam. A spray of cold water condensed the steam, creating a partial vacuum that sucked up more water, thus starting the cycle again. The miner's friend was at best a modest success because the metalwork of that day could not withstand the high steam pressures (8–10 atmospheres) required to drive the pump. Shortly after the beginning of the 20th century, Thomas Newcomen, an engineer, and his assistant John Calley, a plumber, improved Savery's design by introducing a piston that was pushed down by air pressure after the steam was quenched. The Newcomen design did not depend solely on high-pressure steam but instead took advantage of air pressure working against the vacuum. The design was a significant improvement over the miner's friend and quickly spread to mines throughout Europe.

The Newcomen engine had one inherent design flaw: the need for the cylinder to be alternately hot—to receive the steam—and cold—to quench it. An instrument repairman at Glasgow University recognized the inefficiency implicit in such a design. He modified Newcomen's design to provide for a cylinder that was always hot and a separate condensing unit that was always cold. The repairman's name was James Watt. He invented the steam engine.

Social Cost of Coal

The extraction of coal from under the earth's surface to meet the energy needs of an expanding and increasingly industrialized society has exacted a terrible price in suffering and death. In the United States, more than 100,000 lives have been lost in this century as a result of coal extraction. Life was so cheap among miners in the early 19th century that British coroners typically did not even hold an inquest if a corpse was identified as that of a miner.

Working conditions were abysmal until the miners improved their lot through collective bargaining and until laws and regulations were enacted concerning health, safety, and child labor. In the 1800s, a 10-hour day, 6-day work week was not uncommon for adults and children alike. In extreme cases, miners might labor 16 hours per day. The work was hard manual labor involving breaking the coal with picks and loading it with shovels. Children as young as 10 years old were employed for underground work because they could drag loaded buckets of coal through very low gangways, some of which were only a few feet high.

Mine Accidents

The principal causes of mine accidents are

- roof falls;
- gas or dust explosions;

- mishaps with hoists, mine cars, or other machinery; and
- suffocation or asphyxiation.

These collectively account for about 90% of fatalities in underground mines in the United States. Of these causes, roof falls are the most serious, causing about 60% of the deaths. Although major mine disasters may claim hundreds of lives, most of the mining deaths over the years have involved only one or two miners. Roof falls, explosions, and other accidents have occurred so regularly that the cumulative death toll far exceeds that of the relatively few, large-scale catastrophes.

In 1900, the annual death rate among underground miners was 3.5 deaths per thousand miners. Increased concern for mine safety, enactment and enforcement of health and safety regulations, installation of safety devices such as gas detectors, and realization that workers in a relatively safe environment are likely to be more productive have led to significant improvements in working conditions. In recent years, the death rate has been reduced by 85% and is now about 0.5 deaths per thousand miners. Despite this substantial change for the better, miners continue to die, and the long, grim history of underground mining is well summarized by a line from a folk song[6]: "Bone and blood are the price of coal."

Diseases

Miner's Asthma. Even if miners successfully escaped the many perils of work underground, they could still fall prey to diseases, which might take tens of years to manifest themselves. As early as the fourth century B.C., the Hippocratic doctors in Greece noted a chronic shortness of breath among miners. The disease came to be known as *miner's asthma.* During the Middle Ages, it was believed that miner's asthma was a punishment meted out by mountain spirits to the mortals who dared to violate their domain. The true cause was first suggested by one of the most bizarre, colorful, and outrageous characters in the history of chemistry, Philippus Aureolus Paracelsus (born Theophrastus Bombastus von Hohenheim). Paracelsus was a genius or madman, or maybe both, whose teachings in pharmaceutical chemistry and medicine were ahead of their time and outraged his contemporaries of the 16th century. Among other things, Paracelsus suggested the use of mercury compounds for the treatment of syphilis, anticipating by almost 400 years the use of organometallic compounds in this application. Paracelsus correctly noted that miner's asthma is not a punishment for some sin but rather is a result of inhaling dust generated by the mining operation.

Pneumoconiosis. The lung disease caused by the inhalation of irritants is *pneumoconiosis.* Two forms of this disease particularly affect coal miners.

One is *anthracosis*, commonly known as black lung disease, which results from breathing coal dust. The other is *silicosis*, which results from breathing rock dust.

Black Lung Disease. Black lung disease results from the accumulation of coal dust in the small air sacs (the *alveoli*) in the lungs. In the alveoli the coal dust forms sores that eventually are replaced by scar tissue. As scar tissue accumulates, the ability of the lungs to exchange waste carbon dioxide for fresh oxygen is reduced. Furthermore, breathing itself becomes much more labored. A person suffering from black lung disease is readily susceptible to chronic bronchitis and emphysema, which involves actual disintegration of the air sacs.

Silicosis. Silicosis occurs when inhaled particles of silica or siliceous rock are ingested by *macrophages* (cells that attempt to engulf foreign objects in the body). The macrophages tend to collect in the air passages of the lungs and form fibrous clots of tissue that can eventually spread to the openings of the lungs or to the heart. The amount of inhaled silica needed to produce silicosis is usually less than one-fourth the amount of inhaled coal dust necessary to produce black lung disease.

The particle size of coal or silica dust particularly susceptible to inhalation and collection in the lungs is less than 5 micrometers (that is, less than two ten-thousandths of an inch). Unfortunately, much of the dust produced during mining is in this size range. A number of actions are taken to mitigate exposure to dust, including using water sprays to keep dust down, improving ventilation or dust collection, and wearing dust masks. The stone dust used as a precaution against explosions is limestone rather than siliceous rock to avoid having an unfortunate choice between dust explosion or silicosis. Periodic medical examinations can help spot these diseases in their early stages.

Nystagmus. Another serious disease that appears only after many years of working in mines is miners' *nystagmus.* This disease results from the very low light levels in the mine. Its first manifestation is a vision defect that ultimately leads to more serious nervous disorders or even total disability. Improved illumination has virtually wiped out the occurrence of nystagmus in the United States, and it is becoming increasingly rare elsewhere.

Working and Living Conditions

During much of the 19th century and into the 20th century, mine workers were barely better off than serfs. For many miners, life above ground was as harsh as working conditions in the mine. Wages were at a poverty level.

Miners were often paid on the basis of the number of tons of coal they produced. In such a system they received no pay for "dead work" such as erecting timbers or cutting through rock, although such activities were necessary for the continued operation of the mines.

Often the only choice of living conditions was the company town. Miners rented company-owned houses and purchased food and other goods from a company-owned store. Housing was frequently substandard. The mining companies set rents and prices at the company store at levels higher than those prevailing elsewhere. To ensure that the miners did not look elsewhere for better housing or less expensive goods, some mining companies paid the workers with a scrip that could only be used to pay rent or purchase goods at the company store. By extending credit when the mines were not producing or when the miners were occupied with a lot of dead work, the mining company could establish conditions in which the miners were permanently in debt to the company, a situation reflected in the song "Sixteen Tons"[7]:

> *Saint Peter, don't you call me, cause I can't go.*
> *I owe my soul to the company store.*

Labor Relations

Miners have been trying for at least 300 years to organize as a group to obtain improved working and living conditions. In 1662, about 2000 miners petitioned Charles II in an effort to obtain better ventilation in the mines. In the United States, miners made sporadic efforts to organize throughout the 19th century. The most successful of the early efforts was the Knights of Labor, an organization that blended trade unionism with religious fervor and an overlay of secret ritualism somewhat like the Masonic orders.

Labor relations in coal mining have been marked by extraordinary violence on both sides. In the Anthracite fields, for example, some miners, mostly of Irish descent, organized a secret society known as the Molly Maguires. This society was responsible for many shootings and bombings in the 1870s until it was infiltrated by a Pinkerton agent who provided information that led to the conviction and execution of the leaders. The violence was by no means confined to one side. In 1897, a sheriff's posse sympathetic to the mine owners fired on a group of striking miners in Lattimer Mines, Pennsylvania, killing 19 people and wounding 41 others.

The United Mine Workers union was organized in 1890. Labor relations have continued to be volatile and occasionally bloody. However, the lot of miners has steadily improved. Wages have risen, working hours and workweeks have been shortened, health and safety regulations have been enacted, and the horrible excesses of the company towns have ceased.

Despite the significant progress in the betterment of working conditions, underground coal mining remains a dangerous, demanding job.

Surface Mining

The alternative to mining coal underground is mining from the surface. Two general approaches are used for surface mining: strip mining and auger mining. Of these, strip mining is by far more commonly used.

Strip Mining

One requisite for successful underground mining is that the roof rock be solid enough to hold together with some support from timbering or roof bolting. When coal seams lie fairly close to the surface, the rock above them may not be solidly consolidated. This situation makes driving tunnels into the coal virtually impossible because of the continual likelihood of the roof collapsing. The only way in which these coal deposits can be mined is by stripping away the soil or rock lying on top of the coal to expose the coal for extraction. The material that has to be removed to get at the coal is called *overburden.*

The expense and difficulty of operating a strip mine depend in part on the amount of overburden that must be removed. An overburden thickness of about 600 feet represents the maximum that would be economically or practically acceptable in most situations. The total depth of the mine might reach 1500 feet. Another important factor in the economic operation of a strip mine is the *stripping ratio*, which is a measure of relative thickness of overburden to the thickness of the coal. Generally, the maximum stripping ratio of overburden to coal that would be considered for operation is 20 to 1.

Methods. Two general methods of strip mining are used: area mining and contour mining. Area mining is used for level land, and contour mining is used for hilly land.

Area Mining. In area mining, a trench is dug and the overburden is piled to one side in a *spoil ridge.* When the coal has been extracted, a second trench is dug parallel to the first one. The overburden from the second trench is piled as a spoil ridge into the first trench. This process is repeated for successive trenches until the lateral extent of the coal deposit has been worked.

Contour Mining. Contour mining is used when the coal outcrops on a hillside or lies beneath hilly overburden. The overburden is stripped away to uncover the coal. As the overburden is removed, it is dumped on the downhill side of the mining operation. This process forms a structure called a *bench*, which consists of the spoil pile of dumped overburden, the exposed coal seam, and the newly cut "high wall" of rock or soil formed by removal of the overburden. Benches are formed along the hillside as the mining proceeds throughout the coal deposit.

The actual mining procedure consists of only a few steps. The area to be mined is cleared and made reasonably level by bulldozers or scrapers. If the overburden is sufficiently consolidated, explosives may be needed to break it up. Large power shovels or draglines remove the overburden. If necessary, power sweepers can be used to clean the coal surface. Small power shovels or front-end loaders break up the coal and load it into trucks.

Machinery. The types of earth-moving and digging machinery used in strip mining depend on the size of the operation and the nature of the overburden and coal to be removed. A *dragline* consists of a bucket that is mounted on a chain, cable, or rope and is passed on a boom some distance over the ground. The bucket is then dragged along the ground to scoop up the overburden. A second cable or rope raises the filled bucket and dumps it onto the spoil pile or into a truck. *Bucket-wheel excavators* were first introduced for strip mining brown coal. The mechanism consists of a wheel with buckets attached to its circumference. The bucket wheel is mounted vertically on a boom. The boom allows the bucket wheel to be moved up against the material being dug. As the bucket wheel rotates, each bucket cuts, removes, and then dumps the material. The large power shovels are scaled-up versions of the shovels seen around civic works projects or foundation excavations. Some of the largest shovels can remove 300 tons of material in a single bite, a feat equivalent to excavating the basement of a house in one pass.

Labor Relations. Labor conditions in strip mines are significantly different from those in underground mines. Strip mines are not nearly as labor-intensive as underground mines because one worker operating a large earth-moving machine can accomplish more than a team of miners working underground. Although accidents certainly occur in strip mines, strip mines generally are much safer than underground mines. Furthermore, the work is healthier because it is done in the open air rather than in underground passages where dust and gases can accumulate. Although some strip mines have not unionized, those that have unionized experience less volatile labor relations than the unions of the underground mines. Thus, production from strip mines is correspondingly less likely to be interrupted by wildcat strikes. In addition, the amount of coal produced per hour worked in strip mines is about triple the amount produced per hour worked in underground mines.

Production. The proportion of coal produced by strip mining has increased steadily in the last few decades. In 1960, about 30% of the coal mined in the United States was obtained by strip mining. By the mid-1970s, this amount had risen to 50%. Today, more than 60% of our coal comes from strip mines. One factor accounting for this growth is the vast increase in the use of lignite and subbituminous coal that occurred in the 1970s. The relative tranquillity of labor relations in strip mines is also important. Virtually 100% of the lignite and subbituminous coal, about 40% of the anthracite, and about 33% of the bituminous coal are produced by strip mines. The use of strip mining in other countries varies widely. In Great Britain, strip-mined coal accounts for only about 10% of the production. On the other hand, the massive deposits of brown coal in Australia and Germany are mined exclusively by strip mining.

Auger Mining

Auger mining was developed after World War II. Auger mining can be used for coal outcrops on a hillside or in strip mines where the stripping ratio at the high wall has become too great for economic strip mining. Auger mining consists of drilling a series of parallel holes into the coal seam. The augers are 2–7 feet in diameter, and the holes may extend up to 300 feet into the coal. As the auger drills its way further into the coal, the cut coal moves out of the hole and falls onto a conveyor that loads it into a truck.

The attractive features of auger mining are its low cost and simplicity. Digging away overburden, drilling, and blasting are not necessary, and timbers, roof bolts, drainage, and ventilation do not have to be provided. Coal production per hour worked is high. An auger mine can be worked by a crew of three and produce 100,000 tons of coal per year.

Environmental Concerns of Strip Mining

Strip mining, even when practiced by the most conscientious mining companies, is an ugly operation. Strip mining began on a large scale in the Appalachian Mountains in the coal-laden hills of Pennsylvania, West Virginia, and Kentucky. Unscrupulous or uncaring companies blasted and tore away the overburden, extracted the coal, and then moved to other locations, leaving behind useless pits, spoil piles, and high walls. This devastation of a once-beautiful land caused the mental image that most people have of strip mining. In many states it has also led to strict laws governing the reclamation of strip-mined land.

Disruption of the Land

The immediate consequence of strip mining is the inevitable disruption of the land. As the operation proceeds, the mine pits and associated spoil piles eventually cover a large area of land. The deeper the mine, the wider the area is cut open to get at the coal. During the active period of mining, the land can no longer be used as it was prior to mining, whether it was used for agricultural production or for its aesthetic and recreational features. Furthermore, the mining can cause serious environmental problems.

Rainwater can carry away mud from the spoil piles. This mud can eventually wind up in nearby rivers or streams and cause pollution. Some of the material carried away by rainwater is fertile topsoil dislodged by mining. In an extreme case the entire spoil pile can be carried downhill in a devastating mud slide. An extension of the runoff problem is the filling of stream or lake beds with mud and soil. This process is called *aggradation*. In areas where considerable mining activity is occurring, aggradation occurs in streams near the mines and in lakes or reservoirs downstream. Aggradation can alter the environment at the bottom of the water and disrupt the aquatic plant and animal communities. If the stream ultimately flows into a reservoir, then the reservoir is filled with deposited sediments much faster than it would have been by natural runoff in the absence of mining. The filling of the reservoir with sediment reduces its capacity for holding water.

The water draining away from mines can become highly acidic. The origin of the acidity is the conversion of pyrite, one of the most common minerals in coal, to sulfuric acid and iron sulfate. As the highly acidic mine water makes its way into streams, the acid can have a severe effect on local fish populations. It is also very corrosive to anything that might be in the water, such as pumps, pilings, or boat hulls. When the acidity of this water is decreased, either by dilution with other natural water sources or by deliberate water-treatment schemes, the iron sulfate in solution is precipitated as a gelatinous, reddish orange slime in the stream bed or in pipes and other water-handling equipment.

Reclamation Laws

Some observers feel that no other activity involving the land has caused as much destruction of the environment as strip mining. Extreme critics have advocated the outlawing of strip mining, although such a proposal is unlikely to be enacted because strip mines provide more than one-half of our coal and therefore about one-fourth of our electric power. Many of the negative aspects of strip mining can be mitigated through enactment and enforcement of land reclamation laws.

Today, a mining company will have a reclamation plan in hand even before the start of actual mining. The valuable topsoil is carefully removed and stored, not buried under tons of rock, as mining proceeds. When the mining process in an area is completed, reclamation follows, sometimes within only a few months. The high wall and spoil piles are leveled, and the land is returned to its original contour. The topsoil is restored and replanted or reforested. An issue of concern is whether the land should be restored exactly as it was before mining, or whether in some cases the land should be upgraded, that is, turned into more productive agricultural land or sites for public recreation.

The well-known dictum "There's no free lunch" applies to the reclamation of mine land. Currently, reclamation probably accounts for a small percentage of the electric bills of the average household. If reclamation laws are made more stringent, the cost of additional reclamation activities will be reflected in an increased cost of coal, which in turn will be charged to the consumer. Situations may also occur in which reclamation of the disturbed land is nearly impossible, for example, if the original topsoil was too thin or if rainfall is inadequate to reestablish vegetation. Such cases have no scientific solution. Instead, the problem becomes an issue to be decided by society: Is the environmental disruption an acceptable price for the heat, electricity, and fuel provided by the coal?

Notes

[1] Kneeland, F. *Preliminaries of Coal Mining*; McGraw–Hill: New York, 1926.

[2] The hoisting machinery is by no means trivial. Electric hoists may have motors rated at 12,000 kilowatts and may be called on to hoist 10,000 tons per day from hundreds of feet in the earth.

[3] A *squib* is a paper tube filled with slow-burning material at one end and quick-burning powder at the other end. The flame slowly travels to the quick powder, which produces a flame for igniting the black powder.

[4] *Encyclopaedia Britannica*; Encyclopaedia Britannica: Chicago, 1969; Vol. 5, p 971.

[5] Humphry Davy was another outstanding scientist of the 19th century. In a sense, Davy's greatest discovery was Michael Faraday. Davy hired the young Faraday as his assistant, a move that started Faraday on his own brilliant career.

[6] The song is "Ballad of Spring Hill" by Ewan MacColl and Peggy Seeger, written to commemorate the 1958 disaster at the Cumberland Mine, Springhill, Nova Scotia, which took 74 lives.

[7] "Sixteen Tons" was written by Merle Travis. It was recorded in the mid-1950s by Tennessee Ernie Ford and became an extremely popular song.

Chapter 6

Coal from Mine to User

Once the coal is removed from the mine, it must be transported to wherever it will be used. Often the coal must be crushed or ground and treated to remove debris inadvertently mixed with the coal during mining or to remove components, such as sulfur, which are undesirable when the coal is used. The various processes used for handling and treating the coal between mining and marketing are collectively known as *coal preparation.* The operations performed in coal preparation are mechanical, such as crushing coal, or physical, such as removal of impurities by taking advantage of differences in specific gravities. *Coal beneficiation* is a term sometimes used synonymously with coal preparation. However, because beneficiation implies improvement, coal beneficiation refers only to those operations that clean the coal. In some limited applications, beneficiation processes may also include drying or making briquettes.

Transporting Coal

Railroads

Railroads are the most economical method of transportation for long-distance shipment of coal. Railroads handle about two-thirds of the coal used in the United States. Worldwide, the fraction of coal shipped by rail is even higher, accounting for about three-fourths of coal transportation. Roughly one-third of all freight revenues received by railroads come from coal shipments.

A long and close association exists between railroads and the coal industry. The association was more evident in the days of steam locomo-

1171-X/87/0117$06.50/0

tives, when railroads depended on coal mines for fuel. Even with the passing of the steam engine, some railroads are still very closely tied to coal because they depend on coal shipments for much of their income. The plight of railroads in the Northeast provides a good example. Some regional carriers, known collectively as the anthracite roads, were virtually driven into bankruptcy by the sharp decline in coal production and shipment that occurred in the 1950s and 1960s in that region.

The most significant recent development in rail transportation of coal has been the unit train. A *unit train* consists entirely of coal cars and shuttles between a mine and one customer, usually an electric power station. The cars are loaded at the mine, hauled to the customer, unloaded (often by some continuous unloading process), and then hauled directly back to the mine without ever being uncoupled or switched to different trains. In some cases the electric utility even owns the cars. The unit train provides many advantages. It is a more efficient method of shipment than having a few coal cars in a train carrying various types of merchandise because it eliminates the need to break up and reassemble the train in a switchyard. When a mine and a customer commit to the unit-train method, they can improve loading and unloading facilities to make these operations faster and more economical. From the railroad's perspective, the unit train offers the prospect of increasing coal-car use, sometimes by as much as sixfold. The customer may lose some flexibility by committing solely to delivery by unit train but may in the long run find it less expensive than having multiple delivery methods.

Unit trains are not without problems. A unit train might consist of a hundred or more cars having a total load of 10,000 tons of coal. In the West, many small rural towns still have the main rail line passing through the heart of town. One long, slow-moving coal train can badly tie up traffic and other activities. As the use of western coal continues to increase, some towns face the prospect of having dozens of such trains per day significantly disrupting the community.

Trucks

Trucks haul about 20–25% of coal shipments. For small shipments of coal moved over short distances, trucks are the most economical method of transportation. The economic advantage of trucks diminishes as the haulage distances or tonnages increase. However, improvements in the payload capacity of trucks translate into increased competitiveness over long distances.

Barges

The cheapest way of moving coal is by barge. Barges move about the same amount of coal as trucks. The use of this method depends on the mine and the customer being reasonably close to a navigable waterway. A problem for users who rely entirely on barge shipping is the possibility of barges becoming icebound in severe winter weather when the demand for fuel is at its peak.

Pipelines

Coal-slurry pipelines offer an alternative to transportation by trains, trucks, and barges. In essence, pipeline transportation consists of three steps:

- coal is ground and formed into a slurry with a fluid medium (in current practice, the medium is water);
- the slurry is pumped through the pipeline; and
- the slurry is separated to recover the coal at the consumer's end.

The energy crisis of the 1970s brought into discussion many approaches to coal use that seemed like new ideas at the time but in fact were not new at all. Coal pipelines are an example. The first patent on the movement of coal by pipeline was issued in 1891. Some coal was reaching London by pipeline as early as 1914. In the United States, a 108-mile pipeline began operation in 1957; it transported coal from a mine in Cadiz, Ohio, to a power plant in Cleveland. The pipeline operated until 1963 when it became uneconomical in competition with decreasing railroad freight rates. A more successful pipeline is one that transports coal from the Black Mesa mine in Arizona to a power plant in Clark County, Nevada, a distance of 270 miles.

The future use of pipelines remains questionable. Pipelines appear to have an element of illogic in that effort must be put into making a slurry at one end and taking the slurry apart at the other end. This argument is no longer as important as it once was because of progress in the technology of using coal–water slurries directly for combustion or gasification. Another issue is the use of water as the slurry vehicle. Much of the desirable low-sulfur coal that is a candidate for pipeline transportation is in western states. Many of these states have semiarid climates. From the farmers' or ranchers' perspectives, the water being shipped out of state in the pipeline may be even more valuable than the coal. Railroads perceive pipelines as a competitor and have vigorously resisted granting rights-of-way for pipeline construction. In many rural states, railroads and agricultural interests have almost always been at loggerheads. However, when these two powerful groups team together, as in opposition to pipelines, the combination is an extremely potent political force. Such political considerations can force postponement or even cancellation of pipeline projects.

Fluids proposed as alternatives to water for transportation are liquefied petroleum gas, fuel oil, crude oil, methanol, and carbon dioxide. Their advantages are (1) they are less corrosive to pipes than water and (2) they are not a vital agricultural resource. All of these fluids cost more than water; thus, they offer no economic advantage compared with water unless effort is put into separating the slurry and recycling the transport fluid. Also, some of the alternative fluids pose potential environmental problems. Methanol, for example, is both flammable and poisonous and could be a serious hazard in the event of a major pipeline leak.

Despite these problems and objections, pipelines offer the promise of being a very useful method of coal transportation. They can be especially economical when a large supply of coal exists at one end and a consumer who uses a large amount of coal at a reasonably steady rate (many power plants fit this specification) exists at the other end. If the slurry-preparation step can be combined with beneficiation, and if the slurry can be burned or gasified directly by the user without dewatering, pipelines can provide significant technological advantages over traditional transportation methods.

Mine-Mouth Plants

The whole issue of coal transportation can be eliminated by building power plants directly adjacent to mines. Such installations are called *mine-mouth plants.* They account for about 11% of coal consumption. Coal travels from the mine to the plant by conveyor belts or by short-haul trucks. Because the short-haul trucks do not have to use public highways, they can be built to proportions able to handle as much as 250 tons of coal (equivalent to two and one-half rail cars). New power plants, particularly in the West, are often built as mine-mouth plants because electricity can be shipped to consumers more cheaply than coal can be shipped to a power plant located near its customers. The nation's first commercial coal gasification plant, in Beulah, North Dakota, was also built as a mine-mouth plant.

Mine-mouth plants are not a universal remedy. Small industrial installations or domestic users do not use enough coal to justify their own mines. Coal must still be transported and distributed to these small-scale users. Some large-tonnage users, such as coke plants, find it important to blend coals from several mines; thus, they cannot be built as mine-mouth plants. Sites adjacent to the mines must be available for the plants, a condition not always easy to meet in urbanized areas of the eastern United States or Europe. Nevertheless, most likely in the 21st century, the proportion of coal used in mine-mouth plants will increase as more use is made of western coals in new power plants or synthetic fuel plants.

Problems of Storing Coal

Coal is almost invariably subjected to some period of storage between mining and use. The transportation of coal is a form of storage because the coal must remain for a length of time in a railroad car or the hold of a barge or ship. Even mine-mouth plants usually have some provision for coal storage so that they can handle demand when irregularities or short-term disruptions in mine production occur. Other plants may have provisions to store huge quantities of coal to guard against the effects of strikes in the mining or transportation industries or to be able to take advantage of favorable coal prices on the spot market (the market where commodities are bought and sold for immediate delivery and immediate payment).

At first consideration, the storage of coal would seem to be a rather trivial problem. Coal should be able to be handled like other rocks and either be dumped on the ground in a stockpile or loaded into bins or bunkers until needed for use. Unfortunately, several things can happen to coal during storage. As the coal dries, it may break apart into smaller pieces. Mild oxidation of the coal can reduce its heating value and destroy its ability to form coke. In extreme cases, the heat released by mild oxidation raises the temperature enough to cause the coal to catch fire.

Slacking

The degradation of size during storage is known as decrepitation, slacking, or spalling and is especially a problem for low-rank coals. Low-rank coals often have very high moisture contents when they are mined; one North Dakota lignite contains up to 40% moisture. These high moisture contents are usually higher than the normal moisture-holding capacity of coal. As coal is exposed to air it loses moisture; it sometimes loses as much as 30% of its original weight during drying in air. The drying begins immediately when the coal is exposed to air, and extensive loss of moisture causes the coal to shrink. This shrinkage causes mechanical stresses inside the coal particles. These stresses build up to a point at which they can fracture the coal. In extreme cases, slacking can be very extensive in a few days.

Slacking is a problem for two reasons:

1. The formation of small coal particles may result in losses during shipment, handling, or subsequent processing because the particles are susceptible to being blown away.
2. As particle size decreases, the surface area of a given amount of coal increases. This increased surface area increases the chance that the coal will react with oxygen in the air. Such reaction causes loss of heating value and loss of coking properties and also causes spontaneous combustion.

Slacking can be a serious problem for lignites and subbituminous coals and is generally of much less concern for bituminous coals. Anthracites can usually be stored with no problem.

Oxidation

Mild oxidation occurs by the absorption of oxygen from the air. In the early stages of oxidation, the ratio of oxygen to carbon in the coal increases without much change occurring in the ratio of hydrogen to carbon. This result suggests that oxygen is being incorporated chemically in the coal structure without the removal of water or carbon dioxide, which are the products of total combustion of coal. The carbon atoms to which oxygen becomes attached liberate less energy on further oxidation or combustion than do carbon atoms attached only to hydrogen or other carbon atoms. On a macroscopic scale this effect is observed as a reduction in the heating value.

The extent of oxidation and the resulting loss in heating value vary with the properties of the specific coal being stored and the method of storage. In extreme cases, up to 40% of the original heating value can be lost in 6 months; however, in well-consolidated stockpiles where the packing of coal particles minimizes air flowing through the coal, the heating-value loss can be held below 1% per year even over several years. As a rule, the loss of coking properties at such low rates of oxidation is not serious and is of no consequence at all if the coal is being used as a fuel rather than for coke production.

Spontaneous Heating and Combustion

The most serious problem that can be encountered during coal storage is spontaneous heating and combustion. The loss of coal is itself an economic waste, but if a fire actually starts, then property damage, injury, or loss of life can occur. The heating of coal in the hold of a ship, for example, puts the ship and crew in great danger. Unfortunately, the likelihood of spontaneous heating is very difficult to predict. The susceptibility to spontaneous heating depends on the rank, composition, reactivity, particle size, and porosity of the coal. It also depends on the conditions of storage, including the access of air and moisture to the coal pile, the ability for heat to be conducted out of the mass of coal, and the prevailing conditions of temperature, air currents, and humidity.

Several processes may be responsible for the start of heating in coal. Scientists do not generally agree on which process is the most significant. Future research may show that the heating is either a result of several processes operating at once or that one process predominates at a given time, depending on the specific coal and storage conditions. The most straightforward process is the absorption of oxygen from the air onto the coal surface and the subsequent reaction of the absorbed oxygen with the coal. Alternatively, the heating could begin with the oxidation of pyrite. The heat liberated from the pyrite oxidation raises the temperature of the surrounding coal material until it too is able to react with oxygen. Pyrite oxidation has often been suggested as a leading cause of spontaneous heating. However, low-rank coals, which are particularly prone to spontaneous heating, usually contain little pyrite. Another possibility is the heat produced when the coal becomes wet. Heat is liberated when a liquid wets the surface of a solid. This process occurs because the atoms or molecules on the surface of a solid are at a higher energy state than those in the interior. When a new surface is formed, such as by wetting, the molecules that were formerly on the surface are now in the interior and relax to a lower energy state, giving off energy that is observed as heat. This heat of wetting could also contribute to heating the coal pile.

Trouble really starts when the heat liberated from whatever source—oxygen absorption, pyrite oxidation, or wetting—is not conducted out of the coal pile as it is liberated. If the heat stays in the coal, it raises the temperature. Raising the temperature increases the rate at which reactions take place. Increasing the reaction rate increases the rate of heat liberation, thus starting a vicious cycle that leads to the coal catching fire. The ignition temperature depends on the particle size and rank of coal, the amount of oxygen in the air immediately surrounding the coal, and the velocity of the air currents that move fresh oxygen in and conduct heat out.

Limiting the access of air and moisture to the coal prevents or minimizes spontaneous heating. Careful compaction of the coal pile is the best preventative technique. Temperature monitors in the pile can sense the formation of hot spots, which are the first sign of trouble. The first approach to combating hot spots, if the coal is stored in a bin or bunker, is to seal all possible air access routes. If this approach does not work, or if the coal is not in a container, the hot coal has to be dug out and allowed to cool. Sometimes spontaneous heating can be arrested by spraying the coal with water, but this approach can actually be counterproductive when the heat of wetting is contributing to the heating problem.

The storage of coal is a topic that should be as simple as finding room to store the coal. The problems, however, appear to involve some complex coal chemistry and are not well-understood on a molecular scale. These problems are long-standing, are well-known to shippers and users of coal, and have been described on a gross or macroscopic scale. Some investigators have worked on developing empirical methods to predict heating or ignition. Procedures are available for dealing with storage problems at least on an ad hoc basis. However, understanding the problems enough to provide solid approaches to prediction or prevention requires further research in coal chemistry and physics.

Upgrading Coal

Methods for upgrading or beneficiating coal fall into three general categories:

1. removal of minerals physically mixed with or incorporated in the coal;
2. removal of constituents chemically bound to the coal, particularly organic sulfur; and
3. drying, the removal of water.

Coal beneficiation serves several purposes. It can increase the heating value and remove materials that might cause problems when the coal is processed later. For example, in coke making, a large amount of ash tends to dilute the mass of plastic coal and results in a poorer grade of coke. Beneficiation can reduce the amounts of pollutants formed during coal use and thus make the task of controlling emissions of pollutants easier. For example, the removal of pyrite from coal used in power plants can substantially lower the amount of sulfur burned to sulfur oxides and can thus decrease the need for pollution-control equipment at the plant. Beneficiation also offers the potential of making coal easier to handle, store, and ship. Today, the removal of mineral matter is the only category of beneficiation being practiced on a substantial commercial scale. Organic sulfur removal and drying are areas of current research and most likely will become commercial technologies as both the use of coal and the concern for environmental quality increase.

Depending on such factors as the local geological conditions, the thickness of the seam, and the mining conditions used, more than one-fourth of the material brought to the surface during mining may actually be waste. The waste can include shale and other rocks, low-grade coal, coal balls (mineralized nodules of plant remains), and even petrified logs. Because the shipment of waste is not economical, the preparation plants are usually located as close to the mines as possible.

Manual Waste Removal

The simplest procedure for waste separation is removing the waste by hand. In the Anthracite region, coal-preparation plants were called breakers because the coal usually had to be broken down into various size classifications for sale. Until the passage of child-labor laws, boys were commonly hired to watch the coal and pick out shale as the coal moved down the chutes and over the sizing screens in the breaker. Sometimes these breaker boys were not yet in their teens. Being a breaker boy was often the first step in a lifetime of employment in and around the mines.

Even when only a small percentage of the coal is rejected as waste, the enormous tonnage of coal mined leads to waste piles that grow to sizable proportions over the years. The waste is described by a variety of terms that vary from one coal-mining region to another: culm, bone, bony coal, slack, duff, or slag. In Richard Llewellyn's classic novel *How Green Was My Valley*[1], the growing slag heaps provide an ominous background for the narrator's description of the gradual deterioration of the quality of life in his Welsh mining village. If slag piles contain a high proportion of pyrite mixed with combustible material, enough heat may be liberated from the oxidation of the pyrite in air to cause the whole pile to smolder fitfully. The worst tragedy associated with slag occurred, ironically, in Llewellyn's Wales. Steady rains saturated a slag heap near the Merthyr Vale colliery until the wet material suddenly slipped downhill, destroying several homes and the village school in Aberfan and killing 116 children and 28 adults.

Mineral Removal

Procedures for removing mineral impurities rely on the difference in specific gravity between coal and the minerals. The specific gravity (the ratio of the density of a substance to the density of water) of mineral-free coal varies with rank but generally ranges from 1.2 to 1.5. The specific gravities of minerals found associated with coal are much higher. For example, kaolinite (a clay mineral) and quartz each have specific gravities of 2.6, and pyrite has a specific gravity of 5. If crushed coal is placed in a fluid having a specific gravity intermediate between those of coal and minerals, the coal will float and the minerals will sink. Separation of the coal from the minerals can then be effected by using various mechanical systems.

The extent to which coal cleaning is performed depends on coal characteristics and economics. The coal characteristic of most concern is the way in which the minerals are incorporated. Minerals that were deposited in the coal after coalification are concentrated either along bedding planes or in major cracks and fissures in the coal. Removal of the minerals from these coals is fairly easy. In some cases, however, minerals were incorporated as the coal formed instead of being transported or moved in after coal formation. For example, pyrite is sometimes seen growing in cavities that were once plant cells. In these cases, the removal of minerals by simple physical processes is virtually impossible.

The economic concern in coal beneficiation is how to strike a balance between the cost of coal cleaning and either the increased revenues resulting from the greater marketability of the cleaned coal or the cost savings from processing cleaned coal. The problem of sulfur-emission control provides an illustration. When coal is burned in a power plant, the pyrite is converted to iron and sulfur oxides. The sulfur oxides pass through the flue of the boiler and ordinarily would escape to the air. To meet environmental regulations, the utility must install a system to remove the sulfur oxides from the flue gases. A flue-gas desulfurization system requires an investment of capital for its initial purchase and an annual expense for its operation. However, if the power plant burned cleaned coal, a smaller amount of sulfur oxides would be formed and thus a smaller flue-gas desulfurization system would be needed. The savings in capital outlay and annual operating costs for the smaller desulfurization system must be considered in terms of the increased cost of cleaned coal.

How well a particular coal can be cleaned cannot easily be predicted. The only satisfactory approach is to evaluate the coal in the laboratory. The method most commonly used is called the *float–sink test.* Coal is shaken with liquids of various specific gravities. Usually two fractions are formed: the floats and the sinks. Compared with the original coal, the carbonaceous material will be enriched in the floats, and the amount of minerals will be higher in the sinks. The amounts of floats and sinks as a percentage of the original coal, as well as the ash contents of the floats and sinks, are determined. Sometimes a third split, the middlings, forms. The *middlings* is material of about the same specific gravity as the liquid medium used for the test, so the middlings tend to remain in suspension rather than float or sink. The data from these tests allow the coal beneficiation engineer to construct a washability curve for the coal being tested. From the washability curve, the engineer or plant designer can determine what liquid specific gravity would be required to achieve a desired ash level in the floats and what the yields of floats and middlings will be.

A perfect separation between the coal and the minerals is never achieved. A virtually continuous gradation runs from pure coal to coal of increasing mineral content, as well as from rock of decreasing carbon content to pure rock. The optimum cleaning for a particular coal depends on its washability, ash level desired or tolerated in the processing of the coal, and economic trade-offs associated with cleaning to various degrees of ash content.

Coal-Cleaning Processes

Coal-cleaning processes are generally divided into two groups: coarse-coal cleaning and fine-coal cleaning, depending on the particle size of the coal being treated. The division between these categories is somewhat arbitrary, but is usually taken to be at ⅜-inch particles.

Coarse-Coal Cleaning. The principal methods of coarse-coal cleaning are jigs and heavy-media separators. Concentrating tables and cyclones are also used.

Jigs. Jigs (Figure 6.1) are inclined, open-topped, rectangular, or half-cylindrical boxes containing a perforated bottom called a *screen plate.* The raw coal is suspended in water in the jig. A pulsating water flow through the screen plate causes the bed of coal particles to stratify. The alternating expansion and collapse of the coal bed caused by the water flow sorts the

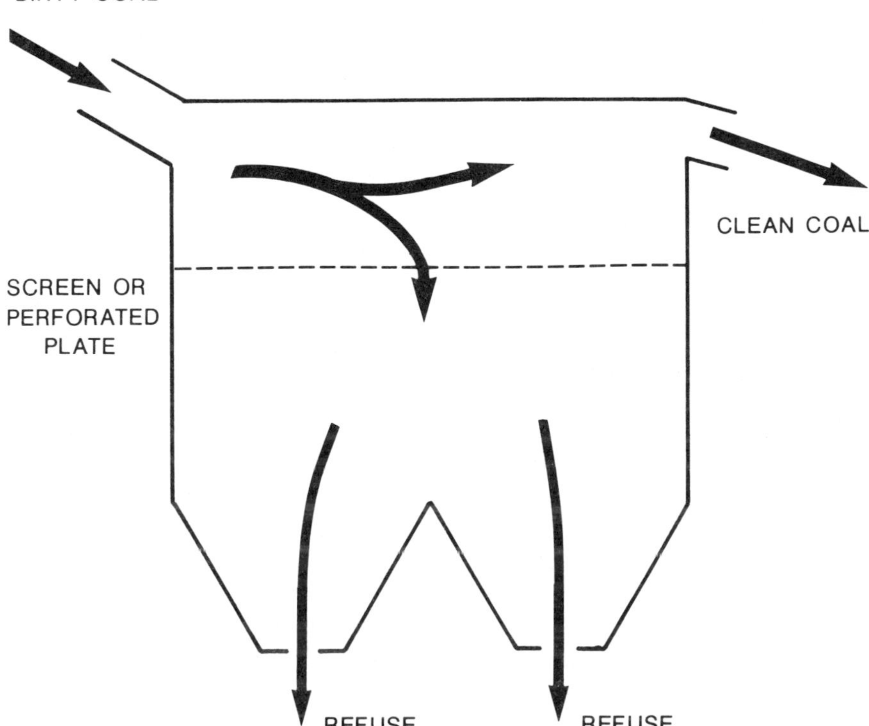

Figure 6.1. In a jig, the coal is supported on a perforated plate through which water flows alternately upward and downward. This flow pattern causes the particles on the plate to stratify by density. The low-density clean coal is swept out of the top; the high-density minerals and dirty coal fall through the plate and out of the bottom.

particles according to specific gravity. Particles of clean coal will have a lower specific gravity than particles of raw coal or minerals. These particles of lower specific gravity rise to the top as the coal bed stratifies and wash out of the jig by overflowing at one end. Meanwhile, the particles of high specific gravity tend to settle on the screen plate and are withdrawn as waste. The behavior of the coal particles is governed largely by specific gravity and only secondarily by particle size. Consequently, jigs are versatile units for cleaning coal in which the size varies widely. Careful screening and size classification prior to jigging are not as important as they are for other coal-cleaning processes.

Because jigs are mechanically very simple devices, they can easily be designed and built to achieve large production rates, up to 700 tons per hour. The use of water as the suspension and separation medium makes jigs economical to operate. Compared with other processes, such as heavy-media separation, jigs do not provide as good a recovery or cleanliness of separation. Nevertheless, the economy of jigs and their versatility in handling a wide size range of coal make them popular units for coal cleaning. A particular application of jigs is the cleaning of coal for use in power-plant boilers.

Heavy-Media Separators. Heavy-media separators clean coal by immersing it in a liquid medium that has a specific gravity intermediate between those of coal and the mineral impurities. The medium most commonly used in coal cleaning is a suspension of magnetite (ferrosoferric oxide, Fe_3O_4) in water. Magnetite has a specific gravity of 5; when mixed with water, it produces suspensions having specific gravities of 1.3–2. Magnetite does not dissolve in water. Using a very fine particle size and a mild agitation of the medium keeps the magnetite in suspension. Sand and ferrosilicon are other materials sometimes used to produce heavy-media suspensions.

Heavy-media separation is not as tolerant of a wide size range of coal as jigging is. The particle size range is set by an appropriate combination of crushing and screening before the cleaning operation. The raw coal feeds continuously into a vessel that circulates the magnetite suspension. Clean coal is removed from the top, and waste is removed from the bottom. The clean coal passes over a screen that allows most of the medium to drain. A second, vibrating screen removes more of the medium. The last traces of medium are removed by a water rinse.

The economical operation of the heavy-media separation process depends on recovering and recycling the medium. Because magnetite is magnetic, any residual fine-coal particles remaining with the magnetite can be discarded by a magnetic separation step. The last bit of suspension can be recovered from the cleaned coal by washing, although washing dilutes that portion of the suspension, which then must be reconcentrated to maintain the appropriate specific gravity.

Heavy-media separation processes are capable of high throughputs of clean coal. By appropriate adjustment of the specific gravity of the medium, very sharp separations of clean coal from refuse can be obtained. Computer-based process-control instrumentation makes it easy to operate large plants with a small work force. The disadvantages of heavy-media separation are the need to limit the size range of the coal being cleaned and the need to maintain and operate a media-recovery system.

Concentrating Tables. A concentrating table (Figure 6.2) consists of a tilted deck on which a series of *riffles* (bars or slats for catching particles) are mounted. A coal–water mixture flows down the table while the table is shaken. The water flow and shaking action combine to stratify the coal

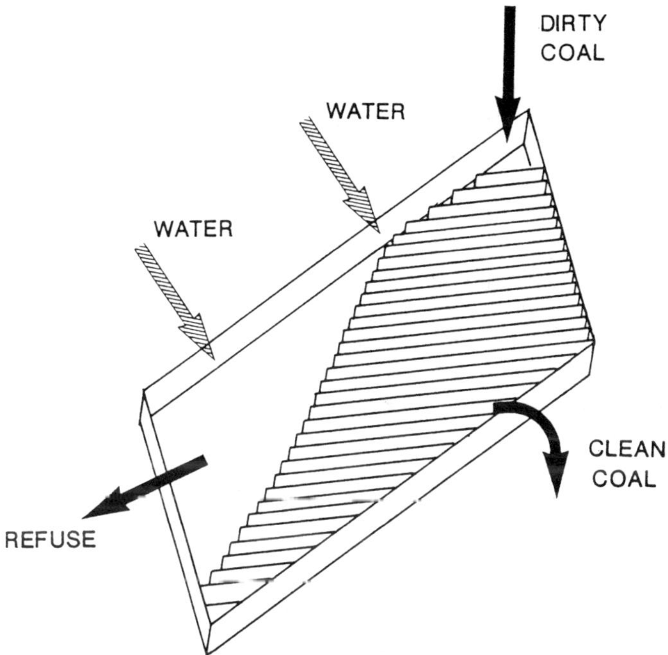

Figure 6.2. The cleaning table uses a cross-current flow of coal and water. The high-density dirty coal and separated minerals flow along the baffles and out one end of the table. Meanwhile, the relatively low-density cleaned coal can flow over the baffles with the water and is removed from another side of the table.

according to specific gravity, similar to what occurs in a jig. The refuse of high specific gravity is held back by the riffles and gradually migrates to one side of the table where it is discarded. At the same time the particles of clean coal stay near the top of the coal–water mixture, flow down the tables over the riffles, and are recovered as the desired product. The shaking motion of the table and swirling of the water make the cleaning table a large-scale analogue of the old miner's pan. The principal difference, aside from the concentrating table being much larger and mechanized, is that the miner panning for gold wanted to retain the material of high specific gravity, that is, gold, and wash away the worthless rock of lower specific gravity.

An advantage of the concentrating table compared with some of the other techniques is that it can handle very fine sizes of coal, down to particles about three-thousandths of an inch. A disadvantage is that a typical concentrating table, which is about 8 feet wide and 16 feet long, can process only about 10 tons of coal per hour. Consequently, a preparation plant using concentrating tables needs to use multiple units. The extra space required adds to the capital investment cost of building the plant.

Cyclones. Cyclones (Figure 6.3) clean coal by taking advantage of the effects of the centrifugal forces set up in a swirling flow of water. Raw coal mixed either with a dense medium such as a magnetite slurry or with water alone is pumped tangentially into the cyclone. As the mixture flows through the cyclone, a vortex flow pattern is established. The material with higher specific gravity moves toward the wall, while the clean coal of lower specific gravity moves toward the axis. A separation is effected by using two discharge points, one for the clean coal and the other for the waste material. Cyclones using a dense medium have a better recovery of clean coal than the hydrocyclones, which rely on water as the fluid. Hydrocyclones do not need the associated medium-recovery system and have a higher processing rate than the dense-medium cyclones. A hydrocyclone 2 feet in diameter can process about 130 tons of coal per hour.

Fine-Coal Cleaning. Jigs, heavy-media separators, concentrating tables, and cyclones might seem at first to be very different approaches to coarse-coal cleaning. Actually, these methods are related by a common physical

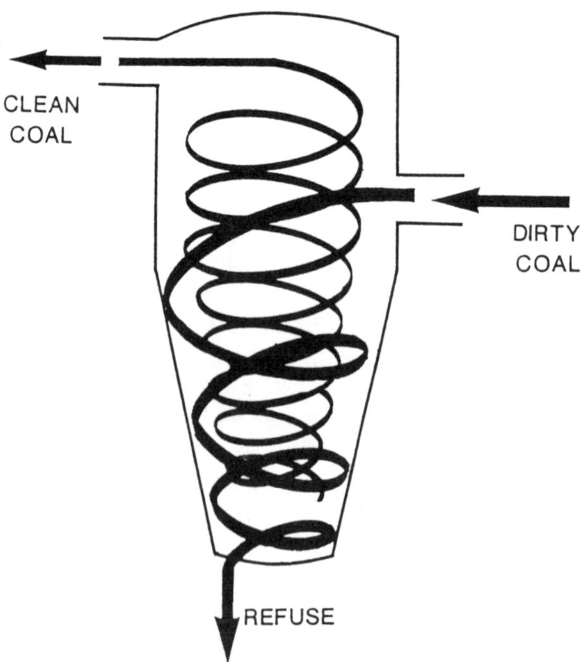

Figure 6.3. A cyclone used for coal cleaning depends on the difference in density between the dirty coal that contains minerals and the clean coal from which many minerals have been removed. The dirty coal has the higher density and travels downward along the walls of the cyclone; the clean coal rides up the vortex in the interior of the cyclone.

principle. These four coal-cleaning methods simply represent different mechanical strategies for achieving a separation of coal and mineral matter based on differences in specific gravities. Other physical properties can also be used as the basis for separation methods. In particular, differences between the surface properties of coal and of minerals are exploited in two techniques for fine-coal cleaning: froth flotation and oil agglomeration.

Froth Flotation. The interaction of water molecules with the surface of a solid depends on the chemical nature of the surface. Nonpolar surfaces show little interaction with, and sometimes even repulsion of, water molecules. Such surfaces are said to be *hydrophobic* (water-fearing). In contrast, polar surfaces readily attract water molecules and are said to be *hydrophilic* (water-loving). The surface of bituminous coal consists largely of aromatic or aliphatic hydrocarbons, is relatively nonpolar, and is hydrophobic. The surface of quartz or clay minerals contains many oxygen atoms and is hydrophilic. Pyrite is generally neither strongly hydrophilic nor hydrophobic.

If air is blown upward through a slurry of finely ground coal in water, air bubbles can attach fairly easily to the hydrophobic surfaces of the coal particles not strongly wetted by the water. The coal particles tend to float along with the air bubbles to the top of the container. The hydrophilic mineral particles are closely surrounded by water molecules and are much less likely to become attached to the air bubbles. Consequently, the mineral particles sink.

This difference in behavior between coal and mineral particles forms the basis of a separation process known as *froth flotation*, so named because the air bubbles beat the coal–water mixture into a froth and the clean coal floats to the surface. A sketch of a flotation cell is shown in Figure 6.4. Skimming off the froth removes the coal; the mineral residue can be removed from the bottom of the flotation unit. Froth flotation is an especially useful process for cleaning fine coal. Currently, it is the only commercial process for cleaning coal in size ranges less than about three-thousandths of an inch.

In practice, the extent of separation can be adjusted by altering the particle size of the raw coal, the amount of coal in the coal–water mixture, and the acidity or alkalinity of the water. Chemicals can be added in small amounts to improve the stability of the froth, improve the sticking of the air bubbles to the coal, or suppress the flotation of undesirable minerals. Because pyrite is neither strongly hydrophilic nor hydrophobic, it may not separate well during a first flotation, but the addition of chemicals that make the pyrite surface hydrophobic and change the coal surface to hydrophilic provides a way to remove pyrite in a second stage of flotation.

Flotation properties also depend on the rank of the coal and the extent to which it has been oxidized during storage and handling. Lignite, which has

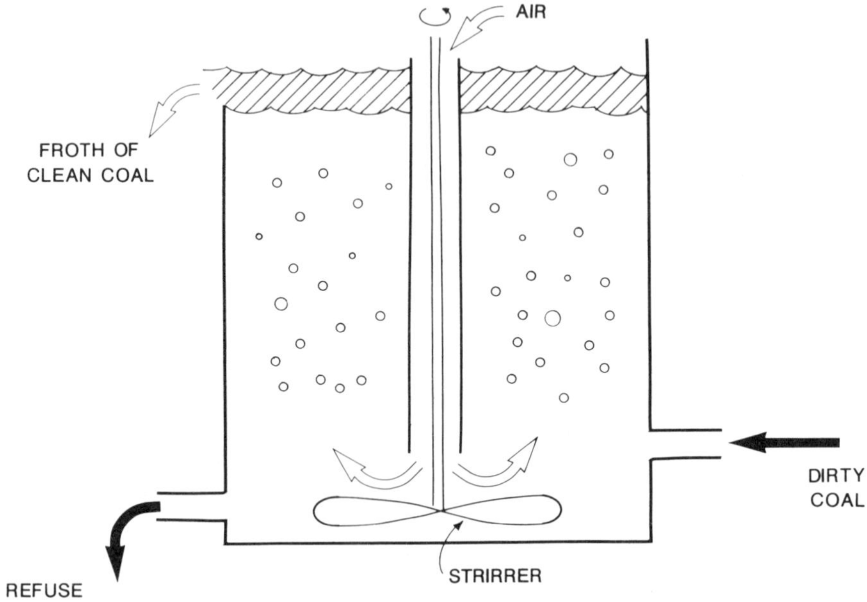

Figure 6.4. The flotation process injects air into a stirred suspension of coal. The hydrophilic coal particles provide sites for the attachment of air bubbles and ride to the top of the vessel. The result is a froth of clean coal.

a high concentration of polar functional groups in the coal structure, has a hydrophilic surface and is the least amenable to flotation of any rank. Low-volatile bituminous coals are the best candidates for flotation. The mild oxidation of coal makes the surface more polar, and hence less hydrophobic, and reduces the ability to float.

Oil Agglomeration. Flotation takes advantage of the coal surface being hydrophobic and thus not able to be easily wetted by water. The *oil agglomeration* process also depends on the hydrophobic nature of the coal surface; however, it also relies on the hydrophobic surface being able to be wetted by a nonpolar liquid such as oil. In fact, oil agglomeration works fairly well with ordinary diesel fuel. In this process a suspension of coal in water is stirred with an oil. The oil preferentially coats the hydrophobic coal surfaces and not the hydrophilic mineral surfaces. The oil-coated particles *agglomerate* (gather into a cluster) and are easily removed by screening.

Like flotation, oil agglomeration works best on finely ground coal. Flotation gives a higher yield of cleaned coal and generally provides more sulfur removal, but agglomeration provides a greater reduction of ash. Oil agglomeration also dewaters coal because the oil absorbed on the surface displaces moisture. Oil agglomeration could potentially be run in combination with a flotation unit to achieve reductions of sulfur, ash, and moisture.

Chemical Coal Cleaning. All of the coal-cleaning methods discussed so far rely on some difference in physical properties for the separation of minerals from coal. The physical cleaning processes are not satisfactory for removing constituents that are chemically bound to the coal structure or that are very intimately associated, for example, minerals growing in the pore cavities of the original woody tissue. Sulfur is one of the prime targets of coal cleaning because of the need to meet air-quality requirements by limiting sulfur oxide emissions from power plants. The removal of comparatively large pieces of pyrite by physical cleaning is fairly easy. However, organic sulfur and fine inclusions of pyrite in pores cannot be removed by physical cleaning.[2]

The development of chemical cleaning methods for the removal of organic sulfur has been the object of much vigorous research during the past 15 years. Many methods have been tried, including oxidizing the sulfur, reducing it, leaching it out with acid, and even letting microorganisms eat it. Although some chemical cleaning processes have shown potential in the laboratory, so far none have become large-scale commercial processes.

The lack of commercial development of a chemical process for organic sulfur removal provides another example of the trade-offs between technology and economics. Physical cleaning processes rely on simple equipment that is easy to build and maintain. Physical processes operate at or near ambient temperature and pressure. Physical processes may use no added chemicals at all, or they may use chemicals that are cheap (such as magnetite) or are used in small amounts (some chemicals added in flotation may be used at a rate of a few ounces of chemical per ton of coal). Some of the proposed chemical cleaning processes use high pressures and temperatures, which add to both the initial investment for the equipment and operating expenses. Chemical processes may also consume larger quantities of chemicals or more expensive ones than physical processes do. Generally, the organic sulfur content of coal is less than 2%. Removing pyrite by physical methods and coping with the rest of the sulfur after combustion is currently more economical than setting up expensive chemical cleaning processes to remove the organic sulfur and residual pyrite prior to combustion.

The chemical cleaning processes that seem most likely to achieve commercial status fall into two categories: hydrodesulfurization and oxidative desulfurization. *Hydrodesulfurization* processes rely on the catalyzed reaction of hydrogen with organic sulfur and pyrite to form hydrogen sulfide. Hydrodesulfurization can do an excellent job of removing organic sulfur. The hydrogen reacts with pyrite to form hydrogen sulfide and ferrous sulfide. Because the ferrous sulfide stays in the coal, hydrodesulfurization removes at best 50% of the pyritic sulfur.[3] *Oxidative desulfurization* converts the organic sulfur compounds to forms such as sulfonates, which are soluble in alkaline solutions. If the reaction medium is already alkaline, these reaction products dissolve out of the coal; alternatively, the products could be removed by a subsequent washing with an alkaline solution. An example of oxidative desulfurization is the Ames process, which treats coal with a dilute solution of sodium carbonate at 300 degrees Fahrenheit and an oxygen pressure of about 14 atmospheres. Although the Ames process has shown promising results, these conditions are far more rigorous than any physical cleaning process. (The trade-off is that the physical processes do not remove the organic sulfur.)

Drying Coal

Another type of coal treatment that results in an upgrading or beneficiation is the removal of water. Water in coal has no adverse environmental effects. However, its presence adds to the weight of the coal and hence to shipping costs. Because water does not participate in thermal reactions, it adds nothing to the heating value of the coal. In extreme cases, such as the gasification of very high moisture coals (about 40%), the heat required to evaporate all the water in the coal is greater than the excess heat generated in the gasification reactions; thus, the process puts itself out.

Removal of water is classified into two categories: dewatering and drying. *Dewatering* is the removal of water added during handling or processing, for example, water added to form slurries for pipeline transport or water acquired during physical cleaning processes. *Drying* is the reduction of the moisture content in the coal below its value as mined. In this book the terms dewatering and drying will be distinguished as just described. However, in much of the coal literature the terms are used interchangeably.

Dewatering Processes

Dewatering processes are usually applied to bituminous coals that have been cleaned. The two mechanical processes generally used are centrifugation and vacuum filtration. In *centrifugation* the wet coal is thrown against a screen by centrifugal forces set up by rapid rotation inside the centrifuge. The water is able to pass through the screen while the coal is held back. *Vacuum filtration* is accomplished by loading the wet coal onto some form of porous medium such as a cloth or screen. A suction is applied to the other side to draw off the water. Thermal-drying and dewatering processes share a common principle: Hot gas, usually produced from combustion, directly contacts the wet coal in some device that allows the gas and coal to flow continuously. The hot gases evaporate the water from the coal, removing the water as steam. Various configurations of thermal equipment are used; a key design feature is the matching of the residence time of the wet coal with the temperature of the hot gases to obtain maximum water removal without thermally degrading the coal itself. A disadvantage of thermal processes is the potential cost. Heating something always adds expense unless process heat or combustion gases, which otherwise would be wasted, can be used.

The Black Mesa pipeline uses mechanical processes for dewatering the slurries. Continued improvements in the direct combustion of coal–water slurries may eventually make the dewatering of slurries unnecessary. The ability to burn coal–water slurries will improve the economics of slurry transport because dewatering equipment will not be needed.

Drying Processes

The drying of low-rank coals presents a situation significantly different from dewatering bituminous coals. First, the moisture content on an as-mined basis is much higher for low-rank coals; one North Dakota lignite has a moisture content of 40%. Second, the cleaning of low-rank coals is not currently practiced on a commercial scale, so dewatering after cleaning is not of concern. The high moisture content of low-rank coals has an impact on every use or potential application of these coals, beginning with transportation. For example, when a unit train of lignite of 33% moisture is loaded, one car in every three is, in effect, a carload of water. Although low-rank coals have very high moisture contents, virtually no commercial-scale drying of these coals is practiced. The only operating installation is at the Sandow plant of Texas Power and Light Company in Rockdale, Texas. The Sandow plant uses thermal drying to treat locally mined lignite, which is then burned on site to produce electric power for an Alcoa aluminum production facility.

Much research has been devoted to the development of drying processes for low-rank coals. Many of the potential processes have been tested on a pilot scale. One of the persistent problems has been that if care is not taken in the drying and subsequent handling, the dried product can reabsorb much of the moisture removed in the drying process. Two drying processes based on the use of steam at high pressures have been proposed for low-rank coals.

The Fleissner process was developed in Austria in the 1920s and is commercially used in Turkey and eastern Europe. The Fleissner process treats low-rank coals with steam at high pressure. The coal particles lose water and shrink. The shrinkage stabilizes the dried particles so that they are more resistant to weathering and to degradation during handling and do not reabsorb the moisture lost during drying. The principal problem with the Fleissner process is the difficulty of getting solid coal into and out of pressure vessels. The Fleissner process operates batch-by-batch rather than continuously. Batch processes operating with relatively inexpensive commodities such as coal are rarely feasible economically.

The Koppelman process is a more recent development that is not currently operating on a commercial scale. The Koppelman process treats a coal–water slurry with steam at 1000 degrees Fahrenheit and 100 atmospheres of pressure. An advantage of the Koppelman process is that a liquid (the coal–water slurry in this case) can be more easily added to a high-pressure reactor than can a solid.

These steam-drying processes change the structure of the coal to make the dried product less able to reabsorb moisture. The reasons for the reduction in moisture reabsorption ability are not yet fully understood, but they probably derive from a combination of collapse of some of the pores by shrinkage of the coal, sealing of pores by tars or residual plant waxes mobilized at the high temperatures of the drying process, and loss of some organic acid groups, making the surface hydrophobic rather than hydrophilic.

A drying process for low-rank coals now under development has the seemingly contradictory name of hot-water drying. In this process, a coal–water slurry is heated to 520–650 degrees Fahrenheit under 55–135 atmospheres of pressure without additional steam. The product, after concentration in a centrifuge, contains 55–60% coal in favorable cases. As in the Fleissner and Koppelman processes, the coal does not readily reabsorb moisture. The goal is to produce stable slurries of about 65% coal. In this case the slurry would have the same proportions of coal and water as the original coal (i.e., 35% water). If a coal containing 35% moisture is slurried with water, the heating value per pound of slurry will be much lower than that of the original coal because of the presence of the slurry water. This reduction in heating value represents an energy penalty for slurrying the coal. However, the heating value of a hot-water-dried slurry of 65% solids is the same as that of the raw coal because no extra water is present. Furthermore, the product is, in a sense, a slurry of the coal in its own water. Transportation of such a slurry from one region to another does not represent a loss of water resources because the liquid medium is only the water that was in the coal in the first place. As combustion engineers develop methods for burning water slurries of low-rank coals, the hot-water drying process may prove to be a significant advance in the transportation and combustion of low-rank coals.

Australian brown coals contain as much as 70% moisture. A mechanical process, *press dewatering*, now under investigation in Australia involves squeezing the coal to force out some of the water. It is similar to methods used for making juices from fruits or vegetables. In fact, apples behave similarly to brown coals during press dewatering. Press dewatering has not yet been tried on coals in the United States; thus, whether or not it will be useful in treating some of the lignites that have high moisture contents remains to be seen.

Discarding Waste

The question of what to do with the discarded material from coal beneficiation remains. Simply piling it up on the ground, the direct approach favored in the early days of the industry, creates hideous eyesores that can be deadly, for example, as in the tragedy at Aberfan. One possible solution is to return the waste material to the mine as fill; another possible solution is to burn it.

The idea of coal beneficiation is to improve the quality of the coal for subsequent use, and today a major use of coal is combustion for electric power generation. At first it seems contradictory to go to the trouble and expense of cleaning coal only to burn the discarded material anyway. However, cleaned coal is needed for the types of boilers used in power stations, in part to help meet air-quality standards. Other types of combustors potentially can take advantage of the waste as a fuel.

Beneficiation waste can be considered as a fuel because of the inability to obtain a perfect separation of coal into mineral-free coal as a product and coal-free minerals as waste. Thus, current beneficiation wastes may contain a large percentage of carbon. The amount of carbon can be much higher in wastes produced years ago when beneficiation processes were not as well-developed as they are today. The wastes, therefore, do contain some potential thermal value.

Fluidized-bed combustors offer a way of using beneficiation waste. In these combustors a bed of some fairly inert material, such as sand, is suspended in a column of air that is blown upward through it. The suspended bed acts in some ways like a fluid, hence the term fluidized bed. The fuel burns in the hot bed. Fluidized-bed combustors are capable of handling a remarkably wide range of fuels. They also operate at rather low temperatures for combustion systems (about 1400–1800 degrees Fahrenheit). The low bed temperature is an advantage in that materials such as limestone or dolomite can be added to the bed to react with and capture the sulfur gases formed during combustion. The in-bed sulfur capture potentially minimizes or eliminates the need for a separate flue-gas desulfurization unit to meet air-quality standards. A fluidized-bed combustor is being tested in Shamokin, Pennsylvania, for burning anthracite culm remaining from the years of vigorous mining activity in the Anthracite fields.

The carbon content of waste from modern beneficiation processes may be too low to provide heat beyond that needed to evaporate the moisture in the waste and keep the combustion going. This situation provides another example of how two technologies, in this case fluidized-bed combustion and coal cleaning, develop in tandem, each mutually supportive of the other. Such a case exists for the preparation of coal–water slurries for drying and transportation, coupled with improved combustion techniques for burning the slurries. Similarly, the possible application of fluidized-bed combustors for using beneficiation waste may affect the beneficiation processes by providing argument for a different separation point and providing a cleaner premium fuel (albeit in smaller quantities) as the main product and a very dirty but still usable fuel as a secondary product.

Notes

[1] Llewellyn, R. *How Green Was My Valley*; Dell Publishing: New York, 1940.

[2] In principle, pyrite inclusions in pores could be removed by physical processes if the coal were ground fine enough to liberate these tiny pieces of pyrite. However, in practice, grinding the coal to this fineness is extremely difficult. Furthermore, such finely divided coal could be hazardous to handle because of the increased likelihood of dust explosions.

[3] This result occurs because pyrite, FeS_2, contains two atoms of sulfur per atom of iron. One of the sulfur atoms becomes converted to hydrogen sulfide and departs, but the other stays behind as ferrous sulfide, FeS. Thus, only one of every two sulfur atoms is removed.

Chapter 7

Heat and Power from Coal

Burning coal is the most straightforward way of using it. The heat from burning coal can be used directly for warmth, cooking, and industrial applications such as heat-treating and shaping metals. Today, these uses of coal account for a small percentage of coal consumption. A much more important use of the heat from coal combustion is to convert water to steam and then use the steam to drive a turbine to produce electricity.

Coal has been used as a source of heat for thousands of years. Because coal seams outcrop on the surface in many parts of the world and because coal is usually easy to ignite, coal may have been used occasionally by prehistoric people. Coal was used in China about 2500 years ago. Coal was known to the ancient Greeks; their word for it, *anthrax*, is the origin of the word *anthracite*. The Romans used coal during their occupation of Gaul and again in Great Britain. Almost nothing is heard of coal after the collapse of the Roman Empire until it was seemingly rediscovered in the 1100s. In 1295, Marco Polo reported that the Chinese were burning a black stone that they dug out of mountains. The use of coal spread throughout Europe and began in North America not long after colonization.

Originally, the use of coal was on a localized basis in the vicinity of the outcrops or primitive mines. However, transportation of coal from mines to a distant market has occurred at least since the 13th century, when coal began to be brought to London by boat from mines in northeastern England. By the 16th century, both the use and shipment of coal began to expand because of severe depletion of supplies of firewood (a situation that is now being repeated in Third World countries). The steady increase in industrialization, development and spread of railroads, and development and increasing use of electric power have resulted in a continually rising consumption of coal during the past 200 years.

Since about 1970, society has become increasingly concerned about the quality of the environment and the effect of the byproducts of coal combustion—sulfur oxides, nitrogen oxides, carbon dioxide, and ash—on the environment. However, people have complained about the smoke and odors from coal fires and have regulated coal burning for as long as records of coal use have been kept. Coal, particularly as a source of electric power, is an absolutely vital component of our modern industrial society and most likely a growing component, but the environmental problems of coal combustion must be recognized and solved.

Small-Scale Uses of Coal

Today, the use of coal for heating homes, small factories, and commercial buildings is just about nonexistent in the United States. The increasing availability of relatively inexpensive natural gas and fuel oil after World War II and the convenience of the all-electric home greatly eroded coal's share of this market. Some observers feel that the decline of coal was hastened by a series of long and bitter postwar strikes that disrupted coal supplies; others feel that shortsighted marketing and pricing policies of the coal industry contributed to the situation. Regardless, the chutes and basement coal bins and the rattle and racket of the monthly coal delivery[1] are largely things of the past, at least in the United States. This situation is not the case, however, in many parts of Europe and Asia.

Burning coal in homes presents two problems:

1. the generally poor thermal efficiency, particularly when coal is used in open fireplaces, and
2. the environmental hazard of smoke and fumes caused by poor combustion.

Fireplaces and Updraft Furnaces

When bituminous coal is burned in a fireplace or an updraft furnace (i.e., a furnace in which the air usually enters beneath the burning coal bed and sweeps upward through the bed and into the flue), the coal begins to decompose in the heat before it actually catches fire. The products of this decomposition are combustible hydrocarbon gases and liquids or tarry hydrocarbons carried out of the combustion region before they can burn. Some of these hydrocarbons recombine to form particles of soot, which are also carried out of the furnace. These products carry away about one-third of the potential heating value of the coal. When coal is burned in an open fireplace, the heat losses are such that only 20%, at best, of the heat available in the coal is actually used to heat the room. The tars and soot that escape combustion contribute to the formation of smoke. Coal smoke also helps initiate the formation of fogs and retards their dissipation. Smoke and soot were regarded with suspicion centuries ago. Elizabeth I would not enter houses in which coal was being burned. She later issued a proclamation forbidding the burning of coal in London while Parliament was in session. The problem of coal smoke and fog becomes particularly acute in regions of high population density where most households and small businesses are burning coal. London, notorious for its pea-soup fogs, became the classic example.

The concerns for heating efficiency and environmental quality may assume a greater importance as some Third World nations switch to coal because of declining supplies of firewood. One approach to increased efficiency is improved design of the heating devices. Water heaters installed in the flue can capture some of the heat of the hot combustion gases, which otherwise would be lost. This improvement can increase the heat recovery of a fireplace to 50%. The addition of transparent, heat-resistant doors boosts the heat recovery to about 65% and still allows the aesthetic pleasures of watching an open fire.

Downdraft Heaters

An alternative design is the downdraft heater. Air enters over the bed of burning coal, moves downward through the coal, passes over the hot bottom surfaces of the furnace, then goes out the flue. This scheme provides ample time for tars and soot to be exposed to high temperatures and burn rather than escape. Downdraft heaters provide good thermal efficiency and almost eliminate smoke. The first design of a downdraft heater was developed in 1680. James Watt revised the design and obtained a patent in 1785. Recently, the design was revised again, in part stimulated by the needs of developing countries.

Fuels

The choice of fuel is also important. Anthracite is the premium coal for domestic use. Although it is hard to ignite, anthracite burns with a hot, clean flame and produces very little smoke and ash. When anthracite is either unavailable or too costly, bituminous coals are used. Because bituminous coals have higher contents of volatile matter than anthracite, bituminous coals can cause problems due to smoke and soot. Methods have been developed for converting bituminous coals into smokeless fuels; some of the leading methods are discussed in the next chapter. Subbituminous coals and lignites are rarely used for domestic heating. Because the heating values of low-rank coals are low compared with bituminous coals or anthracite, larger quantities of low-rank coals have to be burned to achieve the same total heat release.

Burning Coal in Power Plants

The most important use of coal is its combustion to generate steam to produce electricity. For most of us, the use of electricity generated from coal represents the most significant impact that coal has on our daily lives. In the United States, about two-thirds of the coal consumed each year is used by the electric power industry; this coal generates about one-half of our electricity. On a worldwide basis, about 45% of the coal consumed is burned by utilities. Because elsewhere in the world large amounts of coal are still used for direct heating, the percentage of coal used by utilities in other countries is lower than it is in the United States, but the actual amount of coal burned generates two-thirds of the world's electricity. The majority of the power plants use steam turbines in which high-pressure steam, generated by using the hot gases from coal combustion, spins the turbines that drive the generators.

Efficiency

The efficiency of a modern electricity-generating plant based on pulverized-coal firing (described later in this chapter) ranges from 35% to 40%; that is, of every 100 units of energy supplied in the coal delivered to the plant, 35–40 units reach the consumer as electricity. Although a 40% efficiency may seem low, it is nevertheless the result of steady improvements during the past several decades. In the late 1930s, plant efficiencies less than 20% were common. Another method of expressing the conversion of coal to electricity is the *heat rate,* which is a measure of the pounds of coal required to generate 1 kilowatt-hour of electricity. The heat rate depends on the heating value of the coal because it would not be possible to generate the same amount of electricity from equivalent amounts of lignite of 7000 British thermal units per pound (Btu/lb) and anthracite of 14,000 Btu/lb. If a bituminous coal of 12,500 Btu/lb is used as the basis for calculation, the heat rate of a modern, well-maintained power plant is about 0.7 pounds per kilowatt-hour. This value is one-half the typical heat rate before World War II of 1.4 pounds per kilowatt-hour. In practical terms, a power plant having a heat rate of 0.7 pounds per kilowatt-hour generates enough electricity from 1 pound of bituminous coal to operate a 100-watt light bulb for 14.25 hours or to iron clothes with a small steam iron for 1.33 hours.

New designs of power plants on the horizon offer the promise of increasing the efficiency to 45% or lowering the heat rate to about 0.62 pounds per kilowatt-hour. These changes may seem very small at first, but when enough plants of improved efficiency are on-line, these seemingly small changes will be applied against hundreds of millions of tons of coal. One hundred million tons of coal, burned in plants having heat rates of 0.62 pounds per kilowatt-hour instead of 0.7 pounds per kilowatt-hour, will generate an additional 37 billion kilowatt-hours of electricity. This quantity of electricity is almost equal to the annual electric power production of countries such as Argentina, Finland, or South Korea. Alternatively, the electricity now produced from 100 million tons of coal could be generated from 88.5 million tons in plants having the lower heat rate.

Cost

Today, electricity produced from coal is cheaper than that produced from fuel oil or natural gas. Electricity from a nuclear plant is cheaper than coal-generated electricity when the cost of flue-gas desulfurization is included for the coal plant. (If flue-gas desulfurization is not required, coal and nuclear costs are about even.) However, because of the moribund state of the nuclear industry in the United States today, coal can be said to provide the least expensive electricity. Most likely, costs will rise in the future, particularly if environmental quality requirements become more stringent. To comply with tighter environmental regulations, a utility has several options:

- install additional equipment, such as increased-capacity flue-gas desulfurization systems;
- adopt more thorough coal-beneficiation procedures; or,
- for plants in the East, bring in low-sulfur western coal.

All of these strategies result in increased expenses to the utility, which will try like any other business to pass these costs on to consumers.

Combustion Process

The chemical processes involved in the combustion of coal are complex and are still the subject of vigorous research and controversy among coal researchers. Nevertheless, the main steps in the combustion process are generally agreed to be as follows:

1. As soon as a coal particle is introduced to the flame, the heat stimulates the evolution of volatile matter.

2. The evolved gases and vapors catch fire and simultaneously provide a gaseous blanket that retards or prohibits access of oxygen to the surface of the coal.
3. On a time scale of milliseconds, the volatile matter is emitted and consumed, leaving behind a char particle that then ignites and burns.

Chemisorption

The combustion of the char begins with the formation of bonds between oxygen and atoms on the surface of the char, a process called *chemisorption*. The chemical structures formed by chemisorption on the char surface decompose and most likely form carbon monoxide. (Researchers do not agree about whether some carbon dioxide also forms in this step.) The carbon monoxide is rapidly oxidized to carbon dioxide in the gaseous atmosphere around the char particle. The escape of the carbon monoxide exposes fresh surface onto which more oxygen can be chemisorbed to continue the combustion reaction. At the same time, the other components of the coal, principally hydrogen and sulfur, are being converted to their oxides. The mineral matter is left behind, usually altered in composition and mineralogy, as ash.

Heat Liberation

The amount of heat liberated when coal is burned is related to the elemental composition, particularly the amounts of carbon, hydrogen, and oxygen. The nitrogen content of coal is invariably low, usually less than 1.5%. The sulfur content of most coals is low enough not to be of primary importance in affecting the amount of heat generated during combustion. Most of the heat comes from the conversion of carbon to carbon dioxide and the conversion of hydrogen to water vapor. The presence of oxygen in coal lowers the heating value because the oxygen has, in effect, partially oxidized some of the carbon, making that carbon less able to generate heat in the combustion reaction than carbon bonded only to hydrogen or to other carbon atoms. The relative amounts of hydrogen and carbon affect the total amount of heat generated. Comparison of comparable weights of the pure elements shows that hydrogen burning to water vapor produces 3.7 times as much energy as carbon burning to carbon dioxide. The heating value of a good bituminous coal might be about 13,000 Btu/lb. In contrast, gasoline has a heating value of 18,700 Btu/lb, and natural gas has a heating value of 23,400 Btu/lb. Methane, which is the principal component of natural gas, contains 25% hydrogen; an eight-carbon hydrocarbon in gasoline contains 16% hydrogen; and bituminous coal contains about 5% hydrogen.

The relative amounts of volatile matter and fixed carbon help determine the rate of heat liberation, sometimes called the *calorific intensity*. A high proportion of volatile matter helps a coal to ignite easily and burn with a large flame, but the volatile matter does not contribute a great deal to the heat output. In European terminology, coals having a high content of volatile matter are called *flame coals*; these coals burn rapidly. On the other hand, coals high in fixed carbon and therefore low in volatile matter burn relatively slowly but yield a large amount of heat, for example, anthracite. In home furnaces, an anthracite fire can be very difficult to get started, but once the anthracite is burning, the fire will last a long time and produce a great amount of heat.

The molecular components of coal, although altered over millions of years, were originally assembled by plant cells that derived their energy from sunlight. As coal burns, the energy liberated as heat and light is essentially stored solar energy that reached the earth tens or hundreds of millions of years ago.

Coal Selection

All of the traditional coal analyses must be considered when evaluating a particular coal for use as a fuel. The heating value indicates the amount of thermal energy released during combustion of the coal. The volatile matter and fixed carbon results from the proximate analysis indicate the calorific intensity of the combustion process. Moisture has no fuel value, and some of the heat liberated from combustion must be used to evaporate the moisture. The ash determination indicates the amount of solid residue to be handled. Although ash might sometimes be considered the inert portion of the coal, its melting behavior has a major impact on the operability of the combustion system. The ultimate analysis provides data on the sulfur content.

Probably the ideal coal for power generation is a medium- to high-volatile bituminous coal of low sulfur content. Such a coal has a relatively high heating value and enough volatile matter to provide easy and rapid ignition but does not pose serious problems in meeting sulfur-emission regulations. Unfortunately, the supply of such coals is not large. More than one-half of the bituminous coals in the United States have medium to high sulfur contents. Subbituminous coals and lignites generally have sulfur contents around 1%, which is desirably low. However, their heating values are also lower than bituminous coals, and their moisture and ash contents are higher. Thus, shipping the low-rank coals over the long distances from the western mines to eastern power stations involves paying freight charges on an appreciable amount of ash and water. The lower heating value and different ash behavior of low-rank coals make it difficult to switch to subbituminous coal or lignite in a boiler designed to fire bituminous coal. On the other hand, the low-rank coals do serve well for power generation in plants, particularly mine-mouth plants, which have been designed for the specific characteristics of these fuels. Anthracites are low-ash, low-sulfur, high-heating-value coals, but they are so hard to ignite and burn so slowly that they are not good fuels for modern combustion equipment.

Starvation and Excess Air

The amount of oxygen theoretically required for complete combustion of the carbon, hydrogen, and sulfur in coal can be calculated from the ultimate analysis. If less than this theoretical amount of oxygen is available, a situation called *starvation* occurs. The incomplete combustion resulting from starvation means that less than the expected amount of heat will be released and that some of the fuel will be wasted as smoke, soot, or unburned char particles. In actual practice, more than the theoretical amount of oxygen is needed to achieve good combustion. Because the oxygen is actually supplied as air, this extra amount is referred to as *excess air.* A well-designed burner in good condition will operate with 15–20% excess air. Less than optimum situations require more than 20% excess air. The use of excess air contributes to most of the efficiency losses in the boiler. A modern boiler may be 90% efficient[2] in converting the energy of the coal to energy in the steam. Of the lost 10%, 2% is attributed to unburned char or volatile matter, and the rest is the contribution of excess air to the loss of heat in the combustion gases leaving the boiler.

Burning Coal in Power-Plant Boilers

The boiler is the heart of the power plant, the place where the coal is burned to generate steam. The simplest form of boiler is basically a large version of the familiar teakettle. A container of water is heated from the outside; some provision is made for removing the steam. Unfortunately, such a simple system is impractical for any but the smallest industrial facilities needing steam and is certainly impractical for power plants. For the boiler to work, heat must be transferred from the outside into the water. The ability to transfer heat effectively is related to the available surface area across which the heat is moving. Industrial applications may require tens to hundreds of thousands of pounds of steam per hour; power plants may require from hundreds of thousands to millions of pounds per hour. Generating this much steam in a scaled-up version of a teakettle would require a vessel so huge as to be uneconomical, extremely difficult to fabricate, and hard to heat evenly.

Fire-Tube Boilers

Boilers used for large-scale production of steam represent design approaches to increasing the amount of heat-transfer surface while keeping the overall size within reason. One approach is to run a number of hollow tubes through the container of water and allow the hot combustion gases to pass through the tubes. The transfer of heat is not limited by the external dimensions of the boiler but rather by the much larger surface composed of the collective surfaces of the tubes. Such a boiler is called a *fire-tube boiler.* Fire-tube boilers are used in industries requiring relatively small quantities of steam (15,000–20,000 pounds per hour) or very small power plants. The boilers on steam locomotives were fire-tube boilers.

Water-Tube Boilers

Fire-tube boilers are not adequate for high-capacity generation of high-pressure steam. This inadequacy arises from the mechanical problems associated with designing the tubes and the outer container, the *shell,* to withstand the high pressures (ranging to 160 atmospheres) of the system. The solution to this problem is to turn the boiler inside out, in effect, putting the water inside the tubes and letting the combustion gases flow around the outside. This design is called a *water-tube boiler* and is the standard approach used in modern power-plant boilers.

Heat is transferred to the water tubes by two mechanisms: (1) Some of the tubes are suspended directly over the combustion chamber to receive the radiant heat from the combustion reactions. These tubes are in the radiant section of the boiler. (2) Other water tubes are placed in the boiler flue so that the hot gases produced by combustion will flow past them, heating the tubes by convection. This group of tubes is in the convective section or the convection pass. Many boilers also have tubes mounted on the walls of the radiant section; this design is referred to as a *water wall.* Several methods of combustion are used for burning coal in boilers. The principal methods are the following:

- stokers,
- cyclone burners, and
- pulverized-coal burners.

The choice among these methods is based on the size of the boiler (i.e., the amount of steam needed) and to some extent on the kind of coal being burned.

Stokers. *Stokers* are used for small boilers. Several stoker designs have been developed; the principal variation among designs is the relative directions of flow of the coal and combustion air. The fuel and air move in the same direction in underfeed stokers, countercurrently in overfeed stokers (spreader stokers), and crosswise in cross-feed (chain-grate) stokers. Stokers were first developed between 1820 and 1840 and have been continually subjected to minor refinements or improvements. Stokers are the traditional method of firing small boilers having capacities up to 30 megawatts[3] and have been used on units having capacities up to about 70 megawatts. At their largest, stokers have firing rates capable of generating up to 300,000 pounds of steam per hour.

Chain-Grate Stokers. The chain-grate stoker (Figure 7.1) consists of an endless track of grate bars that pass over rotating sprockets at each end, not unlike a conveyor belt. Coal is fed onto one end of the grate from a *hopper.* The coal passes under an *ignition arch,* which is an arch constructed of refractory material heated by the combustion reactions. Heat radiating from the arch ignites the coal passing underneath. The coal bed is about 4 inches deep. The coal burns as it moves along the grate. At the far end of the grate

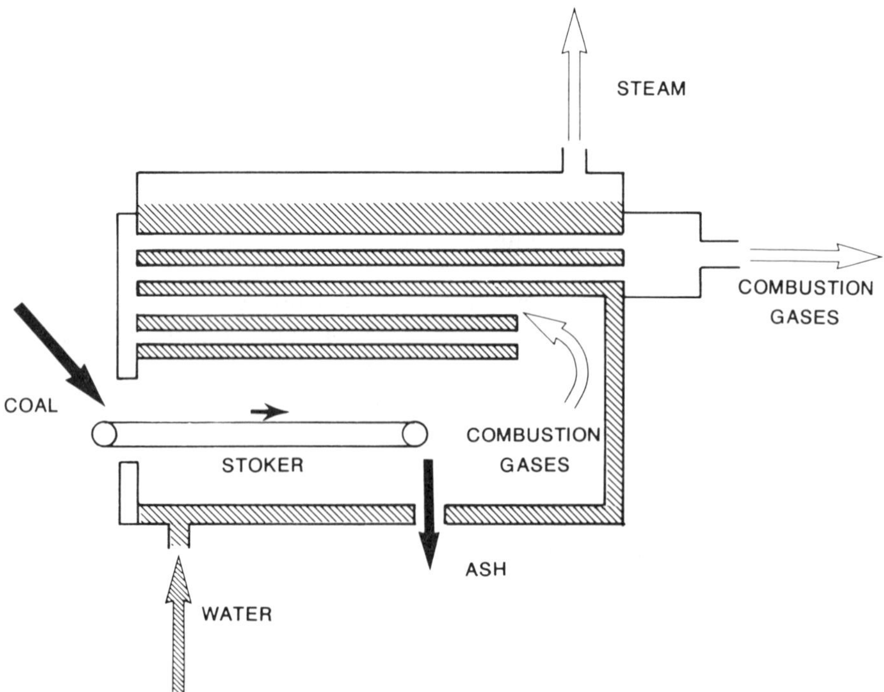

Figure 7.1. Steam can be generated for industrial or small-scale use by using a stoker for burning the coal and a fire-tube boiler. In the fire-tube boiler, the hot combustion gases pass through tubes immersed in the water being boiled. These units do not have enough steam-generating capacity for use in modern power plants.

only ash remains. The ash is dumped as the grate bars turn over the sprocket. The best kinds of coal for chain-grate stokers are coals that do not cake when heated, that is, high-volatile bituminous and subbituminous coals and lignite having high ash fusion temperatures. Coals that swell on heating can clog the air passages in the grate. Ashes of low fusion temperature can soften or partially melt and stick together as clinkers, which will also clog the grate and make ash removal difficult. The low reactivity of anthracites makes them poor candidates for the chain-grate stoker.

Spreader Stokers. In the spreader stoker, coal is fed on top of the bed. A distributor drive on the coal hopper sprays the coal over the bed. The bed is only a few inches deep but consists of three layers: raw coal on top, char undergoing combustion in the middle, and ash on the bottom. Combustion air moves upward through the bed, and ash collects at the bottom. Spreader stokers were very common in old power plants when boilers and generating capacities were small. Virtually any coal can be burned on a spreader stoker.

Cyclone Burners. The *cyclone burner* burns coal that was crushed to pass through a screen having 3/16-inch openings; this burning occurs in a cylindrical, water-cooled combustion chamber. As the name implies, the coal and air mix in a rapid whirling motion and reach combustion temperatures of 3000 degrees Fahrenheit or higher. At these temperatures the coal ash melts to a liquid slag, which runs down the walls and drains from the boiler. The finer coal particles are burned while they are suspended in the swirling air mass. Larger particles, which do not burn as rapidly, are hurled against the wall by centrifugal force, where they stick in the slag until they have burned completely.

Trials of cyclone burners began in the early 1940s; the burners were introduced commercially in 1947 and 1948. In the United States, much of the work leading to commercial success of the cyclone was done by the Babcock and Wilcox Company. The cyclone has several good features, including a very high rate of heat production. It can be designed to use almost any coal from brown coal to low-volatile bituminous coal. The characteristics of the coal ash are very important; the slag must be sticky enough to coat the walls yet fluid enough to run steadily to the slag tap to be drained from the boiler. The moisture content of the coal is also important because excessive amounts of moisture would consume enough heat while being evaporated to lower the temperature generated in the burner; this situation would affect the fluid behavior of the slag. As the temperature decreases, the *viscosity* (a measure of the ability of the fluid to flow) increases.

Researchers originally thought that the cyclone might provide a significant reduction in the emission of ash particles to the air because the ash would be melted to or captured by slag. However, 10–20% of the coal ash is emitted as fine particles that form a visible plume unless emission-control units are installed. Another problem with the cyclone is that it generates large quantities of nitrogen oxide emissions partly because the combustion temperature is so high that nitrogen in the air reacts with oxygen. Despite the positive aspects of the cyclone burner, the associated pollution problems make it unlikely to be a major factor in the power industry in the future.

Pulverized-Coal Burners. In a *pulverized-coal system* the coal is ground so that 65–85% of the particles are smaller than three one-thousandths of an inch and all are smaller than one one-hundredth of an inch. The coal is ground in a pulverizer adjacent to the burners. At this fineness of particle size, a coal–air mixture can be handled almost like a gas. The coal is mixed with *primary air,* the fraction of the total air needed for combustion sufficient to blow the coal mixture into the burners. The *secondary air,* the remaining amount of combustion air, is supplied elsewhere. As the coal particles travel through the burner, they are heated by radiation from the flame; volatile matter is emitted and burned in the primary air.

Unlike in a cyclone, not much of the turbulent swirling needed for complete mixing of the coal and air is provided in the burner. Instead, the turbulence is created in the boiler. Many burner arrangements have been used to achieve this turbulence. In the *opposed-firing method,* burners are mounted directly opposite each other on two walls of the boiler and fire straight at each other. *Tangential firing* uses burners that are aimed along tangents to an imaginary circle located in the center of the boiler. Several burners, each firing along different tangents, collectively provide an intense swirling motion.

The temperature of the combustion reactions reaches 2750 degrees Fahrenheit. Heat is lost by radiation to the water wall and tubes suspended in the radiant section. By the time the gases enter the connective pass, their temperature has dropped to about 1850 degrees Fahrenheit. Most boilers using pulverized-coal firing have been designed to remove the ash as a solid. Therefore, the rate of heat release and the temperature must be controlled to prevent the ash from melting.

The first pulverized-coal burners were introduced in the 1890s for providing heat to cement kilns. Pulverized-coal firing for electric power generation was first used around 1920 by the Milwaukee Electric Railway and Light Company. Since then, it has become the principal method of firing in large power stations. A key advantage of the pulverized-coal system is that the coal requires no mechanical support during its combustion; rather, it burns in suspension. The lack of need for a supporting grate removes restrictions on equipment size. Stoker-fired units typically are limited to plants producing about 10 megawatts of electricity. Pulverized-coal burners can fire plants of 1000-megawatt capacity. The pulverized-coal units operate on about 20% excess air, a factor that helps maintain boiler efficiencies of 80–90%. In contrast, stoker firing may require 50–100% excess air. A pulverized-coal system containing a water-tube boiler is shown in Figure 7.2.

Pulverized-coal firing is not without problems. The need for pulverizers to prepare the fine grind of coal adds to the capital cost of the plant. The power required to operate the pulverizers is the largest contribution to the operating costs of the plant. Although moving the coal directly from the pulverizer into the burner is standard practice, the finely ground coal is nevertheless a potential fire or explosion hazard. The behavior of the ash is also a concern. In addition to the issue mentioned earlier about avoiding slagging, a high percentage of the ash, 60% or more, is blown up the boiler stack as fly ash and must be collected.

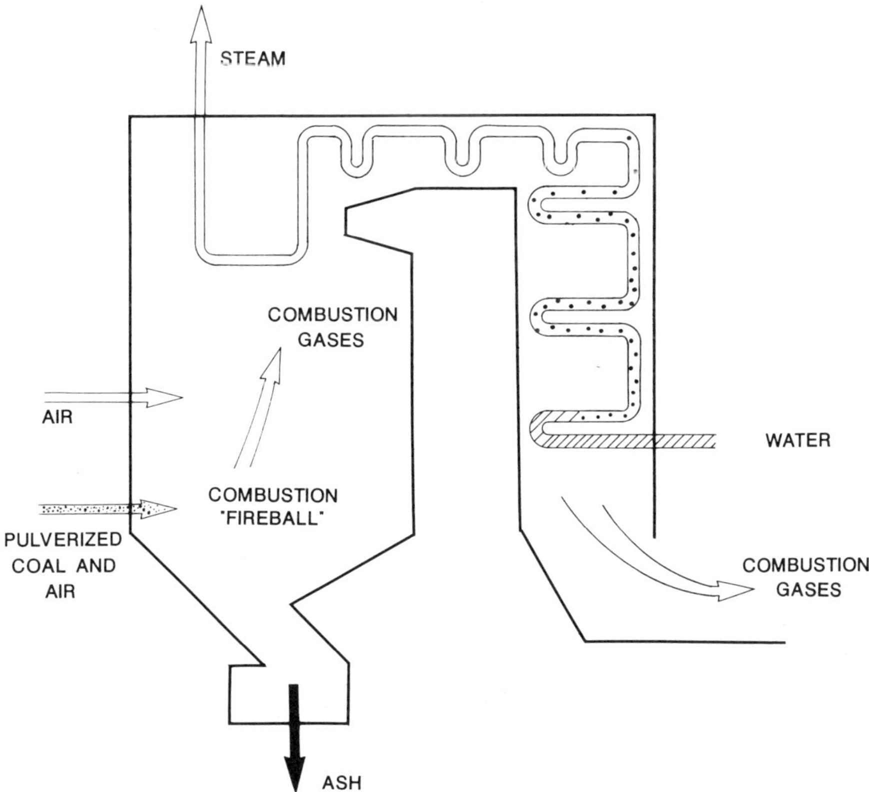

Figure 7.2. The heart of a modern coal-fired power plant is a water-tube boiler. The tubes are heated by the fireball of burning coal particles and also by the hot combustion gases flowing past them. The steam is used in turbines to drive generators. The combustion gases are usually treated to remove ash particles and sulfur oxides.

Ash Behavior in Combustion Systems

The behavior of the ash is a source of great concern in the design and operation of combustion systems. This part of the coal is at least as important as, and perhaps even more important than, the behavior of the carbonaceous part.

Ash, whether formed in the laboratory as part of the ash determination or formed in a combustion unit, is the product of a complex set of reactions experienced by the inorganic constituents of coal during the ashing process. Ash may be the result of several classes of chemical reactions operating on more than a dozen minerals originally contained in the coal. The behavior of ash is very complicated because of the many potential pathways for the minerals to decompose or to react with each other to form new species. These new species might stay solid, partially melt, sinter, completely melt, vaporize, or recrystallize, depending on the prevailing temperature.

The kinds of problems associated with ash can be broadly classified as follows:

- clinkering,
- slagging, and
- fouling.

A simple differentiation is that clinkering occurs on the grate, slagging occurs in the radiant section, and fouling occurs in the convection pass. In practice, such classification is not quite so simple, but it is nevertheless useful for illustrating the types of ash behavior.

Clinkering

Clinkering results from the partial fusion and subsequent agglomeration of ash particles and is a potential problem in stoker-fired units. Because stokers are used only in very small units, clinkering is not a significant problem for the utility industry overall. The principal effect of clinkering is the clogging of air passages in the coal bed. A clinker can form a crust or surface patch that severely limits air access to the coal in the vicinity; this condition causes inefficient combustion and thus less heat generation. A second problem of clinkering is that large, hard clinkers can jam the ash-disposal mechanism.

Slagging

Slagging is the formation of liquid or semiliquid phases. In units such as cyclones that are designed to operate with slag, the concern is to make sure the material remains a liquid. However, in units that are designed to operate without slag, prevention of slag formation is important for three reasons:

1. Ash-disposal mechanisms designed to handle a powdery ash that has never been molten may not be able to cope with large chunks of resolidified slag.
2. Although slag is very hot, it does act as an insulator; thus, slag formation on tubes or water walls reduces the amount of heat transferred and therefore reduces the amount of steam formed.
3. Large deposits of ash that build up on the boiler walls can loosen and fall off because of partial melting of the deposits. The scale of modern power-plant boilers is such that these deposits can be the size of compact cars. When such a mass comes crashing down, it can cause catastrophic damage. Boiler repairs and extra powder purchased to make up for production lost during the repairs can be very costly to the utility.

Fouling

Fouling is the formation of ash deposits on the boiler tubes. Acting as insulators, the deposits reduce heat transfer to the tubes and thus reduce steam generation. In severe cases, the fouling deposits can grow so large that they clog the passageways for gases. The restricted flow of flue gas upsets the combustion process and forces the boiler to be run at a lower rate. Fouling can occur during combustion of either high-rank or low-rank coals. It is of special concern in low-rank coal combustion because it probably is the most serious problem affecting the use of these coals.

Boiler-tube fouling during combustion of low-rank coals illustrates the complexity of ash reactions. The exact sequence of events causing fouling is not yet completely understood on a molecular level, but enough is known to draw some inferences. The process begins with the formation on the tubes of a layer rich in sodium sulfate. Ash particles then begin to accumulate on the front of the tubes (i.e., the side facing the oncoming gases). As the layer of ash particles thickens, it becomes a better insulator; thus, each newly added sublayer of particles cools more slowly. Eventually, this accumulation of ash particles becomes thick enough that the outermost particles have time to undergo a partial melting. Up to this point, the ash deposit is small and poorly consolidated. Once a partially molten material forms, the further accumulation of ash particles results in a continuous, relatively hard solid that can grow very large. In the most extreme cases, deposits can grow so rapidly that the clean boiler can be forced to shut down after only a few days of operation.

Sodium content is the characteristic of low-rank coals that indicates their propensity to foul. The extent of fouling is directly proportional to the sodium content up to 8–10% when reported as sodium oxide in the ash. Beyond these levels fouling is still severe but independent of sodium content. In most low-rank coals, most or all of the sodium is bound to the organic acid groups in the coal structure. As the coal is heated, these acid groups break off and leave the sodium in a mobile state. Simultaneously, the clay minerals in the coal are being dehydrated by the heat; thus, the crystalline lattice of the clay is being opened. The sodium ions have the opportunity to penetrate the opened lattice structure and form new sodium aluminosilicate compounds that have low melting points. These low-melting compounds are the ones that form the matrix of the solid mass of the deposit.

A 1980 study sponsored by the U.S. Department of Energy showed that in 6 months, six power plants burning low-rank coals experienced outages ascribable to ash-related problems, costing the utilities $20 million. Fouling can be a severe problem for plants that must burn high-sodium, low-rank coals. Intense research is aimed at understanding more about the sequence of reactions leading to deposit formation and developing ways to reduce or cope with fouling.

Smoke

Coal is burned to produce hot gases that convert water into steam. These hot gases are the oxides of the constituents of the coal: carbon dioxide, water vapor, sulfur oxides, and nitrogen oxides. This hot gas stream also contains particles of ash. In cases of incomplete combustion, soot and tar vapors may also be in the stream of combustion gases. The release of these substances to the environment is a concern because of their contribution to pollution. This concern is heightened by coal combustion for power generation being the largest use of coal and producing about one-half of our electricity. In addition to the chemical products of combustion, waste heat from a power station is also given off to the environment, a process sometimes called *thermal pollution.* The material we see as smoke is a mixture of soot and condensed tar vapors. Smoke results from incomplete combustion of the volatile matter. Modern combustion equipment, such as cyclones and pulverized-coal units, fortunately provide excellent mixing of coal and air and use of excess air so that smoke is virtually eliminated as an unwanted byproduct. Today, the formation of smoke in a boiler is a symptom that something has gone wrong and usually gets prompt attention from the plant operators.

Smoke formation is a problem with manually or stoker-fired boilers in small power plants and factories and in home stoves and furnaces. In the days when much coal was burned in such ways, smoke was a serious nuisance, especially in areas of high population density. In some places the virtually perpetual cloud of coal smoke reduced the available sunlight by 40%. Around 1910, London experienced an annual soot fall of 426 tons per square mile. Coal burning contributed significantly to the smoky, foggy atmosphere of London. Even that appalling statistic was no match for the city notorious for smoke—Pittsburgh. In 1911, Pittsburgh had soot falling at an annual rate of 1031 tons per square mile.

Smoke was enough of a nuisance because of its dirt, but its long-term effects were much worse than soiled clothing and stained buildings. As long ago as the 17th century, coal smoke was recognized as a health hazard. The first treatise on this topic was written by a London physician, John Evelyn. A particular problem is that sulfur dioxide becomes oxidized in the atmosphere to sulfur trioxide, which absorbs water to form sulfuric acid. The sulfuric acid must condense on the tar or soot particles. When these particles are inhaled, they transport sulfuric acid into sensitive body tissue, particularly in the lungs. At the same time that more than 1000 tons of soot was collecting annually on each square mile of Pittsburgh, the city's death rate from pneumonia was the highest in the world.

Fly Ash

Fly ash is a general term for the solid particles that are entrained in the combustion gases and swept from the boiler with the gases. Fly ash mostly consists of particles of the coal mineral matter that have been altered by the high temperatures of the combustion region. It may contain some carbon as soot or partially combusted char particles. If the fly ash is not removed by control devices, it will be carried out of the boiler stack and released to the environment. Fly ash can be a problem in three ways:

1. Deposition of fly ash can be a nuisance.
2. The smallest particles are respirable and therefore may pose a health hazard.
3. Even when the fly ash is efficiently collected, the question of disposal remains.

Collection Devices

Electrostatic Precipitators. The control method used in most power plants is an *electrostatic precipitator,* which consists of two electrodes between which the combustion gases must pass. The particles of fly ash acquire an electric charge and are attracted to one of the electrodes. In current practice, precipitators are cleaned continually by rapping to dislodge the collected particles into a hopper.

The first electrostatic precipitators came into use in 1923, only 4 years after the introduction of pulverized-coal firing. The requirements for precipitator performance have become increasingly stringent over the years as air-quality standards have become tighter. To achieve a stack discharge that appears clear to the eye, the particles in the gas must be at a level no greater than 200–400 thousandths of an ounce per cubic foot of gas. However, the gas entering the stack has particulate concentrations 100–400 times greater than this level. To achieve a completely clear discharge, the precipitator must operate at efficiencies of 99.0–99.7%. The original units operated at 90% efficiency, and as recently as 1950, 97% was considered good performance.

Meeting emission requirements for fly ash is by no means simple. First, getting any kind of equipment to operate at 99% efficiency is a major engineering feat. The quality of coal used in power stations is becoming increasingly worse because of higher ash contents and ash that is more variable in composition over time. The design and operation of the precipitator must be matched to the characteristics (such as the electric resistivity) of the fly ash being collected, but these characteristics will change as the coal or operation of the boiler changes.

Baghouses. Bags made of fabric can be used to trap the ash particles. These filter bags are usually used in multiples inside a structure called a *baghouse.* A baghouse can offer collection efficiences about as good as those of a precipitor and are used at some power stations.

Cyclone Collectors. *Cyclone collectors* operate on the same principle as equipment used for coal beneficiation. A whirling vortex motion of gas establishes centrifugal forces that separate the comparatively heavy ash particles from the gas. Cyclone collectors are only 70–85% efficient when used on a pulverized-coal-fired boiler. A cyclone-fired boiler produces ash having a larger amount of smaller particles than a pulverized-coal unit produces. A cyclone collector is even less efficient on a cyclone boiler.

Generation and Disposal

Coal-fired electric power generation is thought to contribute about 25% of all the particulate matter in the air. Most of the weight of this particulate matter is due to particles greater than 0.0006 inch (15 micrometers), which will settle under their own weight. Particles smaller than this size can remain suspended for longer times. These particles represent only a small fraction of the mass of fly ash, but because each particle is so minute, the number of particles in this size range is enormous. The particles smaller than 15 micrometers are in the respirable size range and can contribute to health problems.

The amount of fly ash generated each year is about 20 million tons. Much of the fly ash is discarded in various ways, which include returning it to the mine as fill or dumping it in waste ponds or the ocean. Fly ash can easily be viewed as a nuisance to be discarded. However, applications of fly ash can make it a useful and potentially valuable material.

Applications

Concrete. The current use of fly ash is in the manufacture of concrete. Fly ash is useful in this application because of its pozzolanic properties; that is, it can react with lime and water at normal temperatures to produce a cementlike product. Concretes made with pozzolanic materials retain good strength as they age. They are also resistant to acids and salt solutions, a property that is especially desirable on or around highways because the salt used for de-icing forms a saline solution that can damage concrete. Approximately 6% of the fly ash produced in the United States is now used in such applications. Research continues in this area to learn more about making concrete from fly ash and to understand more about the properties of the finished product.

Raw Materials. In the future, fly ash could be a valuable source of raw materials such as aluminum. Today, aluminum is produced from bauxite, a hydrous aluminum oxide. About 90% of the bauxite used in the United States is imported. The major bauxite-mining countries providing supplies to the United States are Australia, Jamaica, Guinea, Surinam, and Guyana. The Soviet Union is also a major producer of bauxite. Except for Australia, these countries could possibly embargo exports of bauxites to the United States either because of domestic instability or as a hostile act. In the 1970s, an embargo on petroleum disrupted life although only a fraction of the total petroleum supply was affected. The U.S. situation with respect to some vital metals is potentially more vulnerable to disruption than the oil situation. Producing aluminum from fly ash is one approach to reducing the dependence on imported ores. However, several problems must be considered:

1. Bauxite typically contains 40–60% aluminum oxide, whereas most fly ashes contain the equivalent of 10–20% aluminum oxide.
2. Aluminum oxide can be separated from impurities in bauxite by simple chemical processes, but most of the aluminum in fly ash is present as aluminosilicates, which are much more

resistant to chemical attack. Consequently, extracting aluminum from fly ash requires more severe and therefore more expensive processing conditions.

3. Fly ash cannot by itself meet the entire demand for aluminum. Processing 20 million tons of fly ash at a nominal 15% aluminum oxide content would provide only about 40% of the national consumption of aluminum.

Rare Metals. Fly ash could also be a source of rare metals. Fly ash contains trace quantities of many elements, some of which are potentially valuable, for example, gallium and germanium, which are used in making semiconductor devices. Although gallium and germanium occur in concentrations measured in parts per million, their high costs (a recent price for high-purity gallium was $11.25 per gram or $319 per ounce) could repay the effort involved in recovering them from fly ash. For example, a medium-size power plant might burn 7000 tons of coal per day and produce about 130,000 tons of ash per year. If the gallium content of that ash is 35 parts per million, then the gallium contained in the ash is worth $1.3 million dollars (if it could be recovered at 99.9999% purity and sold at current market prices).

Sulfur Emissions

Recently, sulfur emissions from combustion have been the focus of much concern. Sulfur occurs in coal in three forms:

- pyrite,
- organically bound sulfur, and
- sulfates.

The sulfates represent a small fraction of the total sulfur in coal and have no significant role in the combustion process itself or in contributing to emissions. During combustion both the organic sulfur and the pyrite are oxidized to sulfur dioxide. Depending on combustion conditions, a small amount of sulfur trioxide may also be formed. Speaking collectively of both of these gaseous products is often convenient, so the formula SO_x (pronounced "socks") is used to represent a mixture of sulfur dioxide and sulfur trioxide.

Fate

The eventual fate of sulfur in coal is a complicated function of the relative amounts of pyrite and organic sulfur, combustion conditions, and prevailing weather conditions. As a rough breakdown, about 15% of the sulfur is retained in the ash. The rest would, in the absence of emission controls, escape to the atmosphere as SO_x. Coal combustion contributes about one-fourth of the sulfur oxides in the atmosphere. Although the details of the chemistry of sulfur dioxide in the atmosphere are not completely known, a portion of the sulfur dioxide is converted to sulfur trioxide and then to sulfuric acid or sulfate salts. About 40% of the sulfur in the coal is deposited as sulfur dioxide by being adsorbed on surfaces (for example, buildings, plants, and cars) or by reacting with them. Another 20% of the sulfur is washed from the air as sulfurous acid in rain or snow; the acid forms as a result of the dissolution of sulfuric acid. The remaining 25% is washed from the air as sulfuric acid or sulfate salts in rain or snow.

Effects

Sulfur pollution is a problem of major concern. Gaseous sulfur compounds are notorious for their obnoxious odors, but a potential odor problem is really the least of worries. Sulfur gases certainly have an effect on health. Usually sulfur dioxide is absorbed in the linings of the nasal passageways and does not reach the lungs. The sulfur dioxide that is converted to sulfuric acid provides a more serious situation because the sulfuric acid coats respirable ash particles that are taken into the lungs and can cause severe breathing problems. Such problems can lead to strain of the heart and lung system. A synergistic effect is seen between sulfur gases and smoke in exacerbating pollution-related health problems.

The deposition of sulfurous and sulfuric acids causes corrosion or decomposition of many materials on which these acids fall. Sulfuric acid is a very strong acid; sulfurous acid is a moderately weak acid but is still corrosive. Iron and steel are fair game for these acids as are many kinds of fabrics, including nylon, cotton, wool, and leather. The acids are also very corrosive to limestone and other carbonates such as marble. Consequently, they can damage buildings and deface monuments, statues, and gravestones that are made of limestone or marble.

Sulfur pollution may also have long-term effects on animal and plant life. The accumulation of acids in natural bodies of water can raise the acidity enough to kill fish and other aquatic life. Similarly, the absorption of acids in soil can harm or kill plants, and the deposition of acids on leaves is also damaging.

Reduction

Dispersion. Because coal now provides about one-half of our electricity and its use in power generation is likely to increase, ways of reducing sulfur emission have to be sought and implemented. The simplest approach is to let the wind disperse the combustion products. Installing very tall stacks on boilers allows the sulfur gases to be dispersed over long distances. This practice serves two purposes:

1. Gases are removed from the immediate neighborhood so that the plant has little effect on the local environment. For a stack 600 feet tall (almost one-eighth of a mile), the concentration of sulfur dioxide at ground level reaches a maximum 3-6 miles away.
2. Gases are mixed with a large volume of air to lower their concentration and reduce their apparent contribution to the pollution wherever they go.

Although the dispersion reduces pollution in the vicinity of the plant and dilutes the gases over long distances, this practice does not reduce the total amount of sulfur introduced to the environment.

Coal Beneficiation and Low-Sulfur Coals. Sulfur emissions can be reduced by burning less sulfur. One method is to clean the coal before it is burned. Coal beneficiation can reduce total sulfur by up to 40%, but it adds to the cost of the fuel. Another method is to switch to a coal having a lower sulfur content. The majority of coal mined in the United States comes from the East, where sulfur contents can be greater than 2.5%. Western coal is typically 1% sulfur. A switch from one type to the other is not simple, however. Burning western coal in eastern or midwestern power plants requires shipping the coal over long distances and adds the cost of transportation to the cost of the fuel. Even more of a problem is that the boiler may not be designed to perform well with western coal. Western coals as a rule have lower heating values than eastern coals. The ash characteristics, particularly the propensity to ash deposition on boiler tubes, may also be quite different for the two coals. Despite the possible problems associated with switching to western coal, it nevertheless is a useful option. Low-sulfur coal having a reasonably high heating value from the Powder River region is now burned in about 20 states.

Flue-Gas Desulfurization. Capturing the sulfur oxides before they can escape to the air provides an alternative to burning less sulfur. This process is known as *flue-gas desulfurization* and occurs in units called *scrubbers*. Scrubbers make use of the sulfur oxides being soluble in water and reactive toward alkaline compounds such as hydroxides or carbonates.

The first flue-gas desulfurization system was used in 1932 at the Battersea power station in London. This system used a spray of water, pumped from the Thames, to dissolve the sulfur oxides. The resulting solution of sulfurous and sulfuric acids was pumped back into the river. Flue-gas desulfurization was used infrequently until the 1960s and then became increasingly common because of the environmental regulations established in the 1970s. Despite the early start in Great Britain, today the United States and Japan lead the world in the application of flue-gas desulfurization, and West Germany is third.

The most common type of scrubber uses a slurry of lime or limestone to react with the sulfur oxides. The product is a sludge containing calcium sulfite and calcium sulfate. For western coals producing ash of high calcium content, an alternative is to use the fly ash instead of lime or limestone. The first scrubbers were plagued with operating problems including corrosion and scaling and in extreme cases even complete plugging. A utility was doing well if the scrubber was running about 40% of the time. Problems have not entirely disappeared, but operability of modern units now exceeds 80%.

Lime or limestone scrubbers are also called *nonregenerative scrubbers* because the scrubbing agent—the lime or limestone—is used once and then is discarded. Nonregenerative scrubbers pose the problem of what to do with the sludge. In a sense, the nonregenerative scrubber transfers the environmental problem from one of air pollution to one of potential land or water pollution. The disposal of the sludge becomes the environmental problem. Many plants impound the sludge in lagoons. As the plants continue to operate, more land area must be devoted to the sludge lagoons. An alternative is to return the sludge to the mine as fill. In either case, water-soluble compounds can be leached from the sludge and contaminate the groundwater. At best, the leaching will produce hard water; at worst, it will introduce harmful trace elements such as arsenic and selenium into the groundwater.

Regenerative scrubbers recycle the active reagent and produce a sulfur-containing material as a byproduct. For example, in the Wellman–Lord process, sodium sulfite reacts with sulfur dioxide to produce sodium bisulfite; the bisulfite subsequently reacts with steam and sodium hydroxide to regenerate the sulfite. Regenerative scrubbers offer several potential advantages. Recycling the sulfur-capture agent eliminates the recurring expense of purchasing this material, except for small amounts to replace losses due to inefficient operation. The sulfur byproduct, which may be elemental sulfur, liquefied sulfur dioxide, or sulfuric acid, depending on the process, can be sold to obtain some income to offset the operating expenses of the scrubber. In the Wellman–Lord process, regeneration of the sodium sulfite makes a concentrated stream of sulfur dioxide that can be used in making sulfuric acid.

Scrubbers and precipitators can significantly help maintain a clean environment. In the late 1930s, before scrubbers were used in the United States, an estimated 2500 tons of sulfurous acid fell in the Chicago area every day. To alleviate such problems, somebody—most likely the utility customer—has to pay. The flue-gas desulfurization system of a 500-

megawatt power station adds 5–7 mils per kilowatt-hour to the cost of electricity (1 mil is one-tenth of a cent). This cost derives from the capital investment for the system, the operating costs, and the labor. In other terms, the flue-gas desulfurization system increases the cost of electricity by about 20–30%. Furthermore, a scrubber does not run by itself. The flue-gas desulfurization system uses 5–15% of the power output of the plant, an amount that otherwise could be sold outside the plant.

Other Methods. These approaches and options for reducing sulfur emissions have been considered in light of conventional (i.e., cyclone or pulverized-coal) combustion systems. A more radical alternative is to change the combustion system itself. One way is to use fluidized-bed combustion, in which a sulfur-capture agent such as limestone is added to the bed to keep most of the sulfur from getting into the gas. Another way is to stop burning coal and instead convert the coal to an essentially sulfur-free gaseous or liquid fuel. Fluidized-bed combustion is becoming a commercial technology and may become a major energy source in the 1990s and the 21st century. In the mid-1980s, synthetic fuels seem less likely to have a significant commercial effect.

Nitrogen Emissions

Nitrogen in coal is oxidized during combustion to a mixture of nitrogen oxides. The principal product is nitric oxide, which accounts for about 95% of the oxides. Most of the rest is nitrogen dioxide, as well as trace amounts of nitrous oxide. Analogous to referring to a mixture of sulfur oxides as SO_x, a mixture of nitrogen oxides is often referred to as NO_x (pronounced "knocks").

Comparison with Sulfur Emissions

A major difference between SO_x and NO_x is that much of the NO_x formed in combustion does not come from the coal but rather from the reaction of nitrogen in the air with oxygen at the high temperatures of the combustion reaction. The oxidation of nitrogen becomes significant at temperatures above 2700 degrees Fahrenheit. The terms *thermal* NO_x and *fuel* NO_x are sometimes used to distinguish between NO_x formed from nitrogen in the air and NO_x formed from nitrogen in the coal, respectively. Up to 80% of the NO_x formed in cyclone burners can be thermal NO_x.

Effects

Nitrogen emissions from coal combustion do not appear to be as significant a problem as sulfur emissions. However, the effect of nitrogen emissions on the environment is not as well-understood as that of sulfur emissions. Nitrogen oxides from coal combustion may represent only 5% of the total nitrogen oxides in air; four times as much is generated from nitrogen-containing fertilizers. Health problems associated with nitrogen oxides are not as great as those associated with sulfur oxides. At high concentrations, nitrogen dioxide does contribute to respiratory disease. Nitric oxide, like carbon monoxide, can block or inhibit the transport of oxygen by hemoglobin. Nitrogen oxides may have a deleterious effect on some plants, particularly grasses, and this effect is worsened when nitrogen and sulfur oxides are both present.

The most notorious role of nitrogen oxides in air pollution is their contribution to smog formation. Nitrogen oxides react with oxygen and hydrocarbons in the presence of light to produce a variety of materials that cause eye and respiratory irritation, for example, organic nitrocompounds and ozone. However, NO_x from coal-fired power plants is not the major cause of smog formation. The major problem is automobile exhausts, particularly in the Los Angeles area, which has become the smog equivalent of the London fogs or the Pittsburgh smoke of the past.

Reduction

Currently, no commercial processes are used for flue-gas denitrogenation. This situation is not critical, however, because variations in the operation of the combustion system can alleviate much of the thermal NO_x formation. The reaction of nitrogen and oxygen becomes appreciable at temperatures greater than 2700 degrees Fahrenheit. However, the reaction can reverse, leading to decomposition of the nitric oxide, if the gases cool slowly from this temperature. High flame temperatures and rapid cooling of the gases provide much less opportunity for the nitric oxide to decompose. Thus, lower flame temperatures should be used, and any nitric oxide that does form should be given time to decompose. Several approaches to accomplishing this result rely on some type of staged combustion. The coal is burned as fuel-rich flame that is later made air-rich flame by delaying the mixing of secondary air. Having only low amounts of excess air is also helpful. Temperature does not seem to affect the formation of fuel NO_x significantly, but this form is generally a small fraction of the total nitrogen oxides.

Various methods are being tested for removal of nitrogen oxides from flue gas. Ordinary scrubbers do not do a good job, mostly because nitric oxide is not very soluble in water. One approach is to react the nitrogen oxides with ammonia in the presence of a catalyst. The products of this reaction are nitrogen and water, both of which are harmless. Research is also underway to develop methods for simultaneous SO_x–NO_x control. For example, a solid that is capable of reacting with sulfur and nitrogen oxides could be injected into the flue. The captured sulfur and nitrogen on the sorbent material could then be collected in a baghouse. Ideally, the sorbent and baghouse combination would provide simultaneous removal of SO_x, NO_x, and particulates.

Acid Rain

Causes

The most controversial aspect of sulfur and nitrogen emissions is acid rain. In even the best of circumstances, pure rain is mildly acidic because some carbon dioxide dissolves in rain as it falls through the atmosphere, producing a pH of about 5.6. The problem of acid rain arises when sulfur and nitrogen oxides remain in the atmosphere long enough to be converted to sulfuric and nitric acids, respectively. When these acids dissolve in rainwater, the pH is lowered drastically, in some cases to about a pH of 3.[4] In parts of New England and the Middle Atlantic states, rain of pH less than 4.5 has been falling for at least 30 years. In some areas of this region, the rain has a pH of 3–4. (To put these numbers in perspective, tomatoes have a pH of about 4, apples have a pH of 3, and vinegar has a pH of 2.)

Effects

A number of noticeable effects have been ascribed to acid rain. Some lakes and streams are becoming so acidic that fish, plants, and other aquatic life can no longer survive. In some lakes in the Adirondack region of New York, acidity increased 30-fold in a decade, and forests are beginning to die. Perhaps the most noticeable case is the serious dying off of trees in the Black Forest in Germany. Buildings and statues made of sandstone, limestone, or marble are being defaced or ruined. The items affected include irreplaceable marble statues from classical Greece, which had survived almost 25 centuries until the increasingly acidic precipitation around Athens began to take its toll. In addition to potential economic losses due to property damage, acid rain can have a more direct effect on humans: Acid water pumped from natural sources through copper plumbing soldered with lead can leach enough copper and lead into the water to present a health hazard.

Something serious is happening; unfortunately, much argument centers around what is causing the problem and what to do about it. There is no clear agreement that the death or stunted growth of trees is a direct effect of acid rain. Furthermore, some water systems undergo cycles of increased or reduced acidity, and in some cases, an ecosystem that is barely maintaining itself on high acidity may be seriously damaged when the acidity naturally fluctuates a bit higher. Nevertheless, a dilute solution of sulfuric and nitric acids continues to fall as rain in many parts of the world, and damage to ecosystems and structures continues to occur.

Role of Coal

The burning of coal contributes only about 25% of the total SO_x in the atmosphere and about 5% of the NO_x. (About one-half of the total amount of these gases actually comes from natural sources such as decaying forest litter and volcanoes.) The options available to reduce the coal contribution to total SO_x and NO_x are the conversion to low-sulfur coals, beneficiation of high-sulfur coals, and installation of equipment to increase flue-gas desulfurization or reduce NO_x formation. All of these options have some associated costs.

Taking Action

Deciding what action to take or advocate involves a trade-off between how much damage we are willing to accept from acid rain (assuming that the effects ascribed to acid rain are indeed caused by it, an assumption that has not been proven) and how much extra we are willing to pay on our electric bills or taxes. Scientists or engineers cannot by themselves provide a direct answer. Chemists can improve our understanding of the reactions involved in SO_x and NO_x formation, capture, and transport in the atmosphere. Ecologists can improve our understanding of the effects of acidification of soil or water systems on plant and animal life. Engineers can design systems to clean coal, control NO_x and SO_x formation during combustion, and capture these materials from the flue gas. Economists can calculate the costs to a utility of installing and using these systems or switching to low-sulfur coal, as well as the additional cost of electricity to the consumer. Ultimately, however, the decisions must be made by society as a whole. What level of damage is acceptable? What costs are acceptable?

Greenhouse Effect

Role of Carbon Dioxide

A gas that is inevitably emitted to the atmosphere during coal combustion is carbon dioxide. Ordinarily, we do not think of carbon dioxide as a pollutant because it occurs naturally in the atmosphere. Carbon dioxide is the fourth most abundant gas in the atmosphere (or fifth most abundant component, if water vapor is counted). However, recent concern has centered around the possibility that increases in the carbon dioxide content of the atmosphere may cause major disruptions to the global climate.

The problem is generally referred to as the *greenhouse effect* because of the way in which a greenhouse retains some of the heat of the sunlight. What we perceive as sunlight reaching the earth is light in the visible region of the spectrum. When the earth radiates heat, the heat energy is given off in the infrared region of the spectrum. Carbon dioxide and water vapor are transparent to visible radiation to allow the sun's energy to reach the earth, but these gases are not transparent to infrared radiation. Therefore, the radiated heat is unable to escape to outer space and instead builds up in the atmosphere.

Climatic Effects

Data obtained during the past 20 years have shown a steady increase of 0.3% per year in the amount of carbon dioxide in the atmosphere. If the increase in carbon dioxide is presumed to lead directly to an increase in atmospheric temperature because of the greenhouse effect, the consequences could be catastrophic. The earth would warm to the point of extensive melting of the polar icecap, which would release enough water to raise the sea level about 200 feet. At first, 200 feet may not sound like much, but this increase would be sufficient to flood many of the world's major cities as well as inundate much low-lying coastal land. Most likely, the productive agricultural regions would shift in accordance with the new climatic conditions. In some cases, productive agricultural regions might shift across national boundaries and exacerbate geopolitical tensions. However, predicting the climatic effects of increased amounts of carbon dioxide in the atmosphere is not this simple.

The amount of carbon dioxide produced annually by combustion is relatively small in terms of the global carbon cycle. Coal combustion introduces about 3 billion tons of carbon as carbon dioxide into the air; the total introduced by all fossil fuels is about 5 billion tons. However, the total amount of carbon dioxide in the atmosphere is equivalent to 700 billion tons of carbon, and the oceans contain about 40 trillion tons of carbon in solution as carbonates and bicarbonates. Each year about 100 billion tons of carbon is exchanged between the atmosphere and the oceans, and another 100 billion tons is exchanged between the atmosphere and the biosphere. Whether or not the oceans would be able to accommodate some of the increased emissions is not known. Also, whether or not increased plant growth would increase the amount of carbon dioxide transferred to the biosphere is not known. (This latter point is made even more complex because large tracts of forested areas, which could be "sinks" for carbon dioxide, are being destroyed every year, particularly in the Amazon Basin.)

A warming of the planet caused by the greenhouse effect could increase the evaporation of water, which in turn would increase cloud formation. Clouds help reflect sunlight back into space. Currently, about

36% of the sunlight approaching the earth is reflected away. If that figure rose to more than 40%, then the additional loss of the sun's heat could cause enough cooling to trigger a new ice age.

Modeling Experiments

The effects of a particular process can be determined in two ways:

1. Do the actual experiment and see the results.
2. Make a mathematical model of the process and calculate the results.

In regard to the greenhouse effect, we are currently performing the experiment on ourselves, although a reliable model would be more desirable. However, in modeling experiments, every factor must be taken into account, and an appropriate mathematical equation must be written to describe the role of each factor. For a simple system, such as a small electric circuit, writing the equations and calculating a voltage or current at some point in the circuit is fairly easy. However, for a planet on which billions of tons of matter are continually being interchanged among its atmosphere, oceans, and plants, the job becomes extremely complex. The equations one selects may not be the most appropriate, and the results may be hard to interpret. Experience tells us that a considerable temperature fluctuation occurs from year to year. For example, the winters of 1977–78 and 1978–79 were unusually severe in many parts of the United States; however, they were followed a few years later by an unusually mild winter. Random fluctuations such as these in data are sometimes called *noise*. Modeling the greenhouse effect produces calculated annual temperature increases that are smaller than the noise. Consequently, determining whether or not the model is accurate over the short term is not possible. If trouble is indeed looming, the solution can be as bad as the problem. For example, carbon dioxide readily dissolves in alkaline solutions; thus, in principle, scrubbing systems could be designed for flue-gas decarbonization. However, this solution poses the problem of what to do with the waste material. One million tons of carbon burned produces 3.5 million tons of carbon dioxide. (One million tons of carbon is the annual consumption of one moderately small power station.) If lime is used to capture this carbon dioxide, then about 8.3 million tons of calcium carbonate is produced and must be discarded. The choice is to have an exceptionally large pile of calcium carbonate or perhaps regenerate the lime and inject the carbon dioxide deep into the oceans. The investment in new equipment would be enormous, and the total electricity put into the power grid would be reduced because additional power would be needed inside the plant to run the extra equipment. The alternative to flue-gas decarbonization is not to make carbon dioxide in the first place. This situation would require a massive and rapid conversion from wood, coal, and petroleum to new energy sources such as solar, geothermal, wind, and possibly nuclear energy. The economic and human dislocations that would be caused by such a conversion are incalculable.

Nobody knows if the greenhouse effect will bring serious consequences in the next century. We do know that we must obtain the most accurate data possible on carbon dioxide emissions, on the fate of carbon dioxide in the environment, and on global weather trends, and we must continually refine the models and their predictions. Because the greenhouse effect potentially affects the whole planet, corrective measures will be effective only through extensive international cooperation on an unprecedented scale.

Look to the Future

Today, the standard approach to coal combustion for electric power generation is a pulverized-coal-fired boiler with SO_x and particulate-control devices. This technology is mature and well-tested. If the characteristics of the coal to be burned are well-understood, then in most cases a pulverized-coal unit can be designed to perform to specifications. Nevertheless, new, innovative approaches to coal combustion may make the pulverized-coal-fired power stations of today obsolete.

Scientists pursue new ways of generating heat or electricity from coal for many reasons:

1. to develop lower cost systems to reduce the capital investment,
2. to achieve more efficient and cheaper control of emissions, particularly SO_x,
3. to improve flexibility in operation to allow efficient response to changing load demand, and
4. to develop systems that can burn lower quality coal or that can accommodate coal of variable quality as premium-quality coal becomes less plentiful.

Fluidized-Bed Combustion

Fluidized-bed combustion is a technology that scientists are developing for the reasons just listed. Fluidized-bed units are just now beginning to penetrate commercial markets. Most likely, more fluidized-bed units will be installed in power stations and small industries in the 1990s and the 21st century.

Principle of Operation. The basic principle of a fluidized bed is that air (or any other gas or liquid) is passed upward through a bed of particles at a velocity sufficient to suspend the particles in the fluid. In such a state, the bed sometimes looks like it is boiling and has the properties of a liquid; if a hole is punched in the side of a fluidized-bed unit, the bed will squirt out through the hole. The vigorous turbulent mixing of the bed provides excellent transfer of heat among the particles and between the particles and the fluid. The rapid motion of the fluid across the surface of the particles provides continuous removal of the reaction products and a continuous new supply of reactant (oxygen, in the case of combustion), a situation that is ideal for promoting reactions between a gas and a solid.

Components. Fluidized beds were first used in the 1940s in the chemical industry in situations in which a reacting mixture was passed through a fluidized bed of catalyst. Engineering development of fluidized beds for coal combustion began in the 1950s. Much of the pioneering work was done in Great Britain, particularly by the National Coal Board and the Central Electricity Generating Board. A fluidized-bed coal combustion system is shown in Figure 7.3.

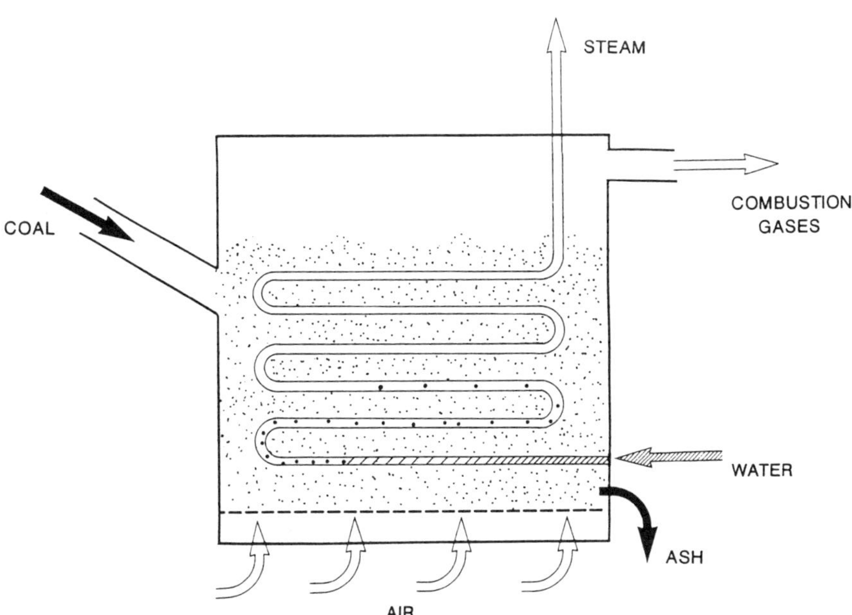

Figure 7.3. A fluidized-bed boiler burns coal in a bed that is suspended in a stream of air. Steam is generated in tubes immersed in the bed. Some designs also have tubes exposed to the hot combustion gases in the space above the bed.

A fluidized-bed combustor consists of a vertical shell containing a distributor plate on the bottom. The bed initially rests on the distributor plate; however, as air is passed through the plate, the bed becomes fluidized above the plate. In some cases the bed can consist entirely of coal, generally pulverized to a particle size less than one-sixteenth of an inch. Alternatively, the bed can include a significant amount of an inert material such as sand. Water tubes are immersed in the bed. Additional tubes can be mounted in the *freeboard* (the space above the bed) or in a convection pass. Particles of ash or unburned coal that are fluidized enough to be swept from the bed are captured in a cyclone.

Combustion Temperature. A key feature of fluidized-bed combustion is the relatively low combustion temperature, typically ranging from 1350 to 1850 degrees Fahrenheit. In contrast, the combustion temperatures in cyclone burners or pulverized-coal units range from 2750 to 3300 degrees Fahrenheit. The low combustion temperature provides several advantages:

1. The formation of thermal NO_x is substantially reduced.
2. The ash fusion behavior is of no consequence because the bed temperature is lower than the lowest ash fusion temperature.
3. The caking properties of the coal are no problem when a bed of sand or similar material is used because the coal particles are far enough apart to avoid agglomeration.

Consequently, the fluidized bed can handle any coal from lignite to bituminous coal exhibiting any degree of caking property and any ash fusion behavior. Fluidized beds can also burn anthracite, although the lower reactivity of anthracite usually requires higher bed temperatures and higher amounts of air than other ranks of coal.

The low bed temperature is also lower than the decomposition temperature of calcium sulfate. This condition means that the SO_x produced in combustion can be captured and retained in the bed. If limestone or dolomite is added to the bed, these materials will react to produce calcium sulfate (and some magnesium sulfate, in the case of dolomite). By keeping the limestone addition controlled so that the molar ratio of calcium to sulfur in the bed ranges from 2–4 to 1, enough of the SO_x can be captured to meet air-quality regulations without the need for an additional flue-gas desulfurization system. Low-rank coals may contain enough calcium and magnesium in the ash so that the ratio of calcium to sulfur is already greater than 2 to 1. These coals, which are usually low-sulfur coals, may be able to meet SO_x emission regulations simply by relying on the natural alkalinity of the ash rather than limestone injection.

Power Reduction. A fluidized-bed combustor can be turned down or idled by reducing the coal feed. The coal feed should be reduced so that the heat generated by combustion just matches that lost in the hot gases and hot solids leaving the unit and any small amounts of steam produced while idling. The unit can even be shut off for several hours. As long as care is taken not to let the bed cool lower than 1100 degrees Fahrenheit, combustion can usually be started again without special measures.

Advantages and Disadvantages. Fluidized beds fulfill most of the reasons for seeking alternatives to current conventional power-station technology. Most likely, some of the new power stations that will be ordered in the 1990s for service after the year 2000 will include fluidized beds. Many of the attributes of fluidized beds also make them attractive for use in industrial installations. The ability to tolerate almost any kind of coal is an advantage for the user who wants to avoid a long-term coal contract. The fluidized bed can also burn coal wastes, such as anthracite culm or byproducts from coal-preparation plants; wood; and even garbage (more fastidiously known as eco-fuel). Meeting SO_x regulations by adding cheap limestone to the bed instead of buying and maintaining a flue-gas desulfurization system is another advantage. Fluidized-bed systems containing a convection pass experience much less ash deposition than pulverized-coal-fired units, a fact that translates into reduced maintenance. Fluidized beds are not competitive in one area: The small boilers have thermal capacities less than 1 megawatt. In these small systems the cost of the control systems becomes an appreciable fraction of the total cost; thus, the fluidized bed is uneconomical.

Cogeneration

A power station incorporating today's standard technology has an overall efficiency of 35–40%. That is, only about 35% of the energy stored in the coal actually becomes useful energy; the rest is wasted. In coal-fired industrial boilers or process-heat units, the coal combustion occurs at very high temperatures, whereas the steam produced is used at much lower temperatures. The temperature difference between the combustion process and the steam also represents wasted energy. In both utility and industrial cases, the efficiency can be improved by making provisions for both the generation of electricity and the capture of some of the available heat. Obtaining electricity and heat energy together is known as *cogeneration.*

Power-Station Uses. By the time a power station incorporating conventional technology has extracted as much energy as possible from the coal, the waste heat is liberated at about 90 degrees Fahrenheit; not much else can be done at that point. However, some of the high-temperature, high-pressure steam used to run the turbines can be withdrawn and used as a heat source instead. This process reduces the amount of electricity generated in the plant, but by effectively using the heat of the high-temperature steam, the overall efficiency can be increased to about 75%. One of the available options for a large power station is to use the extra heat to generate hot water that can then be piped to a town for residential or business heating. This approach is known as *district heating*. In addition to increasing the overall efficiency of the power station, district heating offers the advantage of centralizing the delivery of fuel and the control or collection of wastes. In a district-heating scheme, the hot water system can distribute a great deal of heat over a reasonably large area and achieve minimal heat losses. Smaller power stations have the potential of working directly with a nearby industry or industrial park to supply high-temperature heat that can be used by the industries for generating steam or for process heat.

Industrial Uses. Large industrial facilities generating large quantities of high-temperature steam have the same options for cogeneration as a power station. However, for small industries needing less than 20 megawatts of electricity, the use of cogeneration with standard combustion systems is not cost-effective. Small users have the potential of using coal for generating their own electricity and meeting their heat requirements by adopting alternative technologies. One option is a fluidized-bed combustor. The hot gas leaving the combustor can be cleaned in a cyclone and used in a turbine to drive a generator; the gas that is still hot when it leaves the turbine can be used for making steam. An alternative to the fluidized-bed system is to convert the coal to gas by reacting the coal with steam and air, and then use the coal-derived gas in the turbine. The turbine can be replaced by a diesel engine that uses a coal–water slurry for fuel.

Combined-Cycle Plants

The cogeneration strategy in which the gas turbine generates electricity and the waste heat makes steam can be modified. One modification uses a gas turbine and a steam turbine together to make electricty. A power station using this strategy is called a *combined-cycle plant*. A combined-cycle plant takes advantage of the waste heat leaving the gas turbine. This waste heat is at about the same temperature at which flue gases in a conventional power plant generate steam. In present turbine designs, gases enter at about 2000 degrees Fahrenheit and exit at 950 degrees Fahrenheit. The steam system in a conventional boiler operates between 1050 and 90 degrees Fahrenheit. By adjusting operating conditions, a combined-cycle plant can take in heat at 2000 degrees Fahrenheit and reject waste heat at 90 degrees Fahrenheit. The advantage provided by operating over such a wide temperature range stems from the potential for increasing the efficiency of the plant. For any thermal process, the maximum amount of heat that can possibly be converted to work is given by the difference between the input and exhaust temperatures divided by the input temperature. The maximum possible thermal efficiency for a conventional steam plant is about 64% and for a combined-cycle plant is about 77%.

Types. Two basic types of combined-cycle plants have been developed. One type uses fluidized-bed combustors, operated at high pressure, to generate hot gas for the gas turbine. The other uses coal gasification to make a fuel gas that runs the gas turbine. In either case, the hot gases exiting from the gas turbine can make the steam needed for the steam-turbine cycle. Gasifiers have the advantage because at the higher inlet temperature, the gas turbine is more efficient. Fluidized-bed combustors are limited to maximum temperatures of about 1750 degrees Fahrenheit because higher temperatures can result in problems from ash fusing in the bed. In addition, although fluidized-bed combustors permit substantial sulfur capture in the bed, gasifiers can achieve more than 99% sulfur removal for the product gas.

On the basis of today's turbine technology, which limits inlet temperatures to about 2000 degrees Fahrenheit, a combined-cycle plant using a gasifier is slightly more efficient than a standard pulverized-coal-fired power station using flue-gas desulfurization—approximately 40% versus 37%, respectively. Substantial research is now being devoted to developing new turbines that can operate at higher inlet temperatures, up to about 2600–2700 degrees Fahrenheit. The efficiency of a combined-cycle plant using these turbines is expected to range from 43% to 45%; this efficiency provides a significant advantage over today's standard technology. A block flow diagram of a combined-cycle plant based on gasification is shown in Figure 7.4.

Initial Tests. The first tests of a combined-cycle plant were conducted in Luren, West Germany, in 1972. This plant employed five Lurgi gasifiers (discussed more fully in Chapter 10) having coal capacities of 240–360 tons per day. Lurgi gasifiers have the disadvantage of producing a tar byproduct and a "dirty" gas containing particulate matter, moisture, and tar droplets. The tar and gas have caused problems, but the combined-cycle system itself has worked well. In the United States, a combined-cycle plant is operating at Southern California Edison's Cool Water station near Barstow in the Mojave Desert. This power plant uses a Texaco gasifier rated at 1000 tons of coal per day. The Texaco gasifier does not have the tar problems that the Lurgi gasifiers have. The results from Cool Water have been very encouraging. The successful operation of Cool Water and further advances in turbine design allowing higher inlet temperatures most likely will lead to the building

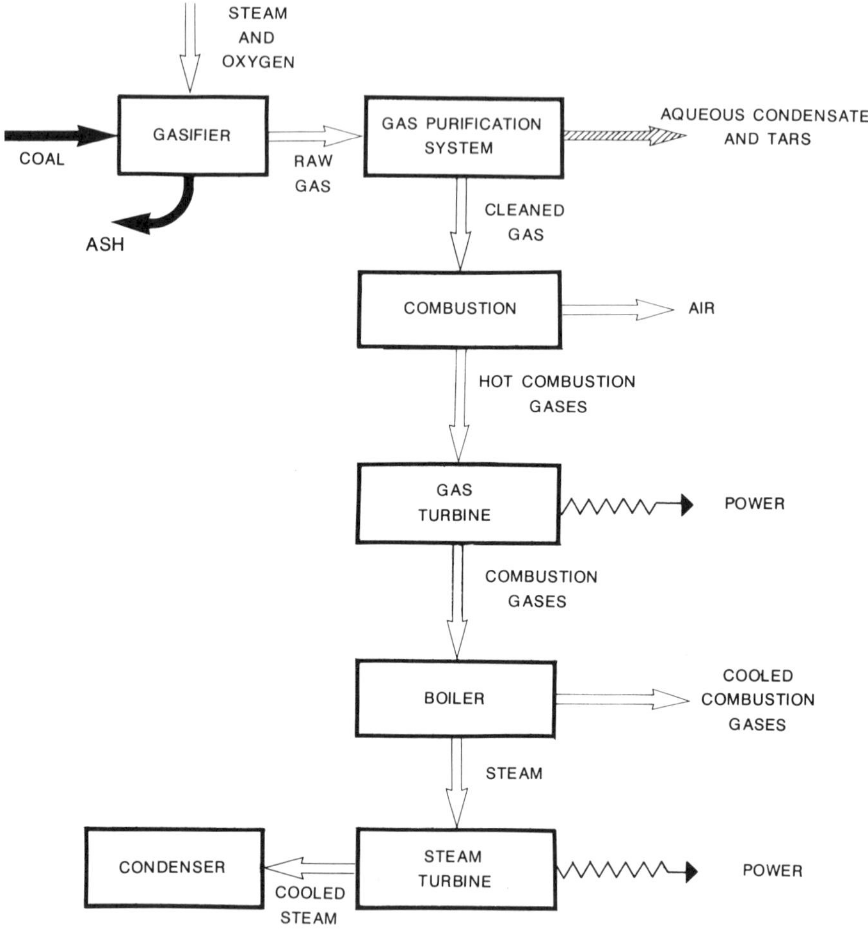

Figure 7.4. One approach to a combined-cycle plant is to burn the gas produced by a gasifier and then use the hot combustion gases to drive a gas turbine. The gases leaving the gas turbine are still hot and can be used to generate steam to drive a steam turbine in the second cycle of the plant.

of commercial-scale power stations using the combined-cycle concept with Texaco gasifiers in the 1990s or the beginning of the 21st century.

Notes

[1] P. G. Wodehouse, in his short story "Stiff Upper Lip, Jeeves", once likened a particularly horrible sound to "the delivery of several tons of coal through the roof of a conservatory."

[2] This figure refers to the efficiency of the boiler only. The efficiency of the whole power plant is much lower, ranging from 35% to 40%. Most of the difference between the 90% efficiency in the boiler and the 40% overall efficiency is accounted for by losses in the turbines.

[3] This unit is a measure of the thermal energy generated by the boiler. At typical plant efficiencies, a power plant with one such boiler can produce about 10 megawatts of electricity. Such a power plant is very small by today's standards.

[4] The pH scale is logarithmic, which means that a change of two pH units, for example, from 5 to 3, represents a 100-fold increase in acidity.

Chapter 8

Carbonizing Coal

The Industrial Revolution changed the course of Western civilization and forever altered the place of people in the world. The changes brought about during the Industrial Revolution fall into four categories:

1. the substitution of machinery for human brawn and labor,
2. the centralization of work into mills or factories instead of rudimentary manufacturing done in the home,
3. the substitution of steam power for animals, and
4. the development of new processes for making iron and steel.

The availability of large deposits of high-quality coal was an essential factor in the Industrial Revolution, first in Great Britain and soon after in Belgium, Germany, and the United States. Although resources were available, new ways for using them had to be developed. One of the principal technical developments that made the Industrial Revolution possible was the inexpensive large-scale production of iron.

Iron-Ore Smelting

Iron occupies a unique position among the metals. Along with its alloys, iron is the most versatile material available to engineers; it is relatively easy to form into shapes, strong and durable in use, and somewhat resistant to corrosion. Other metals, such as copper, tin, and lead, are much easier to extract from their ores but lack the desirable engineering properties of iron.

1171–X/87/0177$06.00/0

Yet other metals, for example, aluminum and titanium, are superior to iron for some technical applications but could not be produced by the relatively primitive metallurgical techniques of the late 1600s. The primacy of iron was described by Rudyard Kipling[1]:

> *"Gold is for the mistress—silver for the maid—*
> *Copper for the craftsman cunning at his trade."*
> *"Good!" said the Baron, sitting in his hall,*
> *"But Iron—Cold Iron—is master of them all."*

Kipling was thinking of bayonets and sword blades as cold iron; however, for metallurgists and engineers as well as for militarists, iron is "master of them all" when compared with other metals.

Unlike gold, silver, and some copper deposits, iron does not normally occur in nature in a metallic state. Instead, it occurs in a variety of ores that are oxides or carbonates: hematite, magnetite, limonite, taconite, and siderite. To make the metal, the ores must be *smelted,* that is, decomposed by being heated with a reducing agent that combines with the oxygen to liberate the metal. The various forms of carbon are the cheapest and easiest reducing agents to use. Iron-ore smelting began around 3000 B.C. Large-scale iron working began in Asia Minor in 1500 B.C. Charcoal, made by incomplete combustion of wood, was the main carbon source for smelting for the next 30 centuries.

Use of Charcoal

As population increased in the 1600s, industrial expansion occurred. The increasing industrialization made serious demands on the available supplies of wood for making charcoal. The devastation of England's forests became so severe that Parliament acted to restrict the building of new iron works. Smelting a ton of iron required the charcoal produced from 3 acres of trees. The amount of available wood became a severe constraint on the expansion of iron making. Thus, the relatively small iron production of the 1600s was not sustained for long on charcoal. A modern technology based on charcoal-smelted iron would be impossible. In 1984, the United States produced about 100 million tons of steel. If charcoal had been used, 300 million acres of trees would have been required, that is, one-third of the total forest lands of the United States. This example demonstrates how limitations of energy supply can put constraints on industrialization. Iron ore was abundant and accessible; however, iron making could not expand and iron goods could not become readily available until a new fuel was found to replace wood charcoal, which was in short supply.

Use of Coal

Coal was considered as a potential replacement for charcoal. The town of Dudley in Worcestershire was conveniently situated atop deposits of the three ingredients for iron making:

- coal,
- iron ore, and
- limestone.

(Limestone is used to form a molten slag with impurities in the ore.) Coal mining had begun in this region under the reign of Edward I in the late 13th century, and this coal was part of the richest seam in England. In 1609, Dud Dudley, an illegitimate son of the local lord, experimented with using coal in iron furnaces and obtained a patent on the process in 1621.

Substituting coal for charcoal was not particularly successful. When coal is heated, its complex molecular structure is broken down into a variety of smaller molecules. Some of these molecules can introduce impurities into the newly liberated metal and result in a poor grade of iron. Sulfur in the coal can form hydrogen sulfide as the coal decomposes; the hydrogen sulfide reacts to make iron sulfide in the metal. When a cast piece of iron is shrinking as it cools in a mold, the iron sulfide can set up planes of weakness in the metal, which sometimes cause the metal to crack. Iron sulfide also makes the hot metal more likely to break while it is being rolled into shape, and finished pieces are weaker and less ductile than the purer metal. Because of such problems, little enthusiasm was seen for using coal instead of charcoal, even though wood supplies were dwindling.

Coal as a replacement for charcoal was also causing problems in other industries. English brewers used charcoal as a fuel for drying malt. When coal was used instead, some of the volatile species released during the heating of the coal were absorbed by the malt and ruined the flavor of the beer. Someone eventually tried heating the coal just enough to drive off the gases and then used the remaining solid as fuel. To perform this technique without burning the coal, air has to be kept away from the coal. This process of heating coal in the absence of air is called *carbonizing*. When some bituminous coals are carbonized, the remaining solid is a hard, porous mass of high carbon content called *coke*.

Use of Coke

While the iron makers and brewers were trying to cope with a switch from charcoal to coal, Abraham Darby, a brass founder in Bristol, was considering applying brass-casting methods to iron. Although brass was an important component of commerce and technology at the beginning of the 18th century, the prospect of using iron instead of brass in items was appealing because ironware could be much cheaper and thus would sell more widely than comparable articles of brass. Around 1710, Darby moved his operations to the Shropshire coal fields near the town of Coalbrookdale and began successfully using coke, rather than coal, to smelt iron ore. The development of coke smelting in Darby's works is attributed to an assistant, John Thomas. Virtually nothing is known about John Thomas, yet he, James Watt, and a handful of other inventive geniuses were instrumental in starting the Industrial Revolution.

The new process of smelting with coke was less costly and used less fuel per ton of iron produced than charcoal smelting. Nevertheless, several decades passed before coke smelting became fully accepted. Coke contains more ash, sulfur, and phosphorus than charcoal. The sulfur and phosphorus find their way into the iron as impurities, and the ash adds to the slag that must be removed from the furnace. In the early 1700s, iron made with coke was brittle and hard to shape. This problem occurred partly because coke needs a greater oxygen supply than charcoal to burn fiercely, and the iron furnaces of the time provided an insufficient amount of air. In 1735, Darby and his son, who was also named Abraham, invented a tall blast furnace with a powerful draft. The Darbys added other modifications to the process, and after the mid-18th century, coke became the standard fuel for iron smelting.

Thus, the development of coke smelting resulted from the depletion of wood resources; the availability of iron ore, coal, and limestone; and inventive and inquisitive people such as the Darbys and John Thomas. However, coke smelting might not have developed if the coal that was buried in the Worcestershire and Shropshire seams had not been a certain type.

Properties. To be used successfully in iron smelting, a coke must have the right properties:

- good mechanical strength,
- high porosity, and
- low sulfur content.

Not all coals yield a satisfactory coke when carbonized. If small samples of the various ranks of coal are heated in laboratory crucibles, a wide range of behaviors is observed. When this experiment is done under standardized conditions, the amount of coke produced is a measure of the *free-swelling index*. The free-swelling index ranges from 1 to 9; generally, a good coking coal has a free-swelling index greater than 4; these coals are usually medium- or low-volatile bituminous coals.

Formation. On a molecular scale, the sequence of changes from coal to coke is not well-understood because of the complex and heterogeneous molecular structure of coal. When many bituminous coals are heated, they begin to soften. For the prime coking coals, the softening begins before the heat begins to break apart the coal structure into gaseous decomposition products. As the gases evolve, they pass through the plastic mass of coal, leaving behind pores. During this plastic stage, aliphatic carbon–carbon bonds or carbon–oxygen bonds between aromatic ring systems are broken. The products of low molecular weight may escape as simple gases, such as methane, or form a complex mixture of compounds that subsequently condenses as a tar. The large, high-molecular-weight, aromatic ring systems remaining recombine and solidify as coke.

Graphitizing Behavior. A common strategy in coal research is to simulate the behavior of coal by using compounds of simpler and known structure. Coking has been studied by thermally decomposing a large variety of organic compounds. In general, these compounds fall into two categories: graphitizing and nongraphitizing; the name implies the tendency to produce a solid product having a structure similar to graphite. *Nongraphitizing* materials are low in hydrogen and rich in oxygen, as are many coals having high contents of volatile matter. They produce a rigid, disordered structure when heated. On the other hand, materials rich in hydrogen and low in oxygen tend to undergo an alignment of molecular layers in the plastic stage. The alignment of layers during the plastic stage is *graphitizing* behavior; this behavior is typical of a good coking coal. The structure of graphite is shown in Figure 8.1.

Coking Coals. The chemical analysis of a coal does not by itself indicate whether or not a coal will be a good coking coal. Ideally, the coking behavior should be evaluated in a small-scale test. However, some aspects of coal composition are useful guides for selecting candidate coals. The coal should be low in ash; a large amount of mineral matter in the coal can dilute the plastic stage and thus interfere with formation of the coke. Also, virtually all of the ash in the coal ends up in the coke. In the blast furnace, this ash melts along with the limestone and mineral impurities in the iron ore. As the ash content of the coke increases, more heat must be wasted for melting it and larger amounts of slag must be handled. Good coking coals should be low in sulfur and phosphorus to keep these impurities from getting into the iron. Coking behavior is determined by how softening, volatile evolution, and resolidification actually occur under the conditions of temperature and time used in coke production.

Finding coal that has all of these desirable properties is becoming increasingly difficult. Today, two or three coals are often blended together to produce a coke of appropriate composition, strength, and porosity.

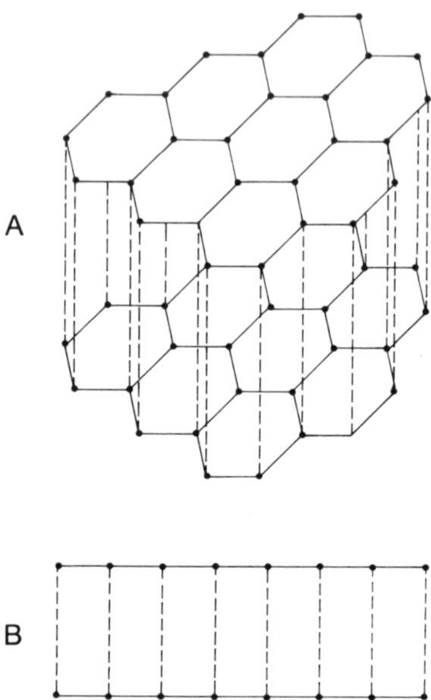

Figure 8.1. Viewed from above (A), the structure of graphite appears to be infinite sheets of carbon atoms in hexagonal rings. A side view (B) shows that the atoms in a sheet are closer to each other than to the atoms in the next sheet. Graphite is, in principle, the ultimate end product of coalification.

Although scientists have attempted to predict the coking behavior of a blend of coals from the properties of the individual coals, a laboratory or small-scale plant test on the actual mixture is the only reliable method.

How Coke Is Made

Mound Process

The first approach to making coke on a large scale was a rough adaptation of the method of making charcoal. Coal was piled on the ground, and horizontal and vertical passageways left in the mound were filled with wood. The wood was ignited, and the heat from the burning wood began carbonizing the coal. The volatile gases and tars released during carbonization also burned, so gradually the fire and heat spread throughout the whole mound of coal. As the fire and heat spread, the mound was carbonized. When the process was judged to be finished, the hot coke was smothered with dirt.

This process did produce some usable coke; however, because some of the coal burned rather than carbonized, the yields of coke were low. Once the mound of coal was set on fire, the process could not be controlled.

Therefore, obtaining a uniform degree of carbonization throughout the pile was not possible. The quality of the coke produced varied considerably. The open-air burning and carbonizing meant that the process also depended on the weather. A mound of coal could take anywhere from 5 to 8 days to carbonize, depending on the prevailing weather conditions. Unburned carbonization tars that escaped into the local surroundings were vile and dirty.

This *mound process* was first used in the United States in Fayette County, Pennsylvania, around 1810. The Cambria Iron Company worked on improving the process, but too many disadvantages were found. By the mid-1800s, a much superior method was used.

Beehive Coke Ovens

The new process involved heating coal in an oven made of clay, stone, or some similar heat-resistant refractory material. These coke ovens were hemispherical and resembled gigantic beehives. The similarity was such that they became universally known as *beehive coke ovens.* A beehive oven was a round structure about 10 feet in diameter having a domed top 7–8 feet high. The oven was loaded with 5–8 tons of coal, which made a *charge* (quantity of fuel) about 2 feet deep. The top of the coal charge was heated by heat reflected from the hot roof and walls of the oven. As carbonization began, enough air was admitted to the oven to burn the combustible volatile matter; this burning supplied more heat. Coking proceeded from the top of the charge down and took 2–4 days to complete. Some of the fixed carbon in the coke also burned. The mass of carbonizing coal expanded so that the finished coke was about 6 inches deeper than the original charge of coal. When coking was complete, the coke was quenched with a water spray and raked out of the oven. The oven was then recharged with a fresh batch of coal. The coking and quenching operations caused vertical cracks in the charge so that the pieces of coke raked out of the oven were about as long as the bed was deep, that is, about 2.5–3 feet. The beehive process made about 1350 pounds of coke per ton of coal.

The beehive coke oven was the standard of the industry for about 50 years, from the mid-19th century to the beginning of the 20th century. It produced an excellent grade of coke if coal of good quality was selected for coking. However, the seeds of the beehive oven's demise were being sown at virtually the same time it was coming into accepted practice.

Byproduct Coke Ovens

Background. In the middle of the 19th century, research on the applications of coal byproducts was providing many new and useful materials. In 1820, Charles Macintosh, a Scottish inventor, immortalized his name by using a naphtha solvent for preparing a solution of rubber that could be used to waterproof cloth. This invention gave the world the waterproof raincoat. In 1856, a young student at the Royal College at London, William Perkin, prepared the first synthetic dye. Nine years later Joseph Lister, then a professor of surgery at Glasgow University, revolutionized medicine by using a spray of phenol (carbolic acid) to drastically reduce the risk of infection during surgery. Organic chemists were learning that phenol, naphtha, and aniline could be derived as useful byproducts from the carbonization of coal. Coal could be a source of many beneficial chemicals, if only they could be recovered and separated for use. However, the whole essence of the beehive process was that the carbonization products were burned to supply heat for the coking reactions.

Although various improvements and modifications were made to the beehive ovens in the late 1800s, the process could not be modified to allow for significant recovery of byproducts. A new design of the coke oven was needed: one that could be heated by a method other than burning the volatile matter and one that could collect and condense the vapors, oils, and tars produced during carbonization. Several inventors, notably Adolph Koppers, developed variations of this basic idea, which came to be known as *byproduct coke ovens.*

The development of the byproduct coke oven is an example of how the changing value, either utilitarian or economic, of byproducts can provide the impetus for a major revision of a process. The byproduct oven dramatically changed the nature of the coke industry. The first 12 byproduct ovens in the United States were erected in 1893. The ovens were actually constructed to obtain ammonia; the coke in this case was the byproduct. Their output in 1893 was a minuscule fraction of the total national coke production.

Conversion of the coke industry from beehive to byproduct ovens was boosted by World War I. Byproduct ammonia, benzene, and toluene were in high demand for conversion to explosives. Coal tar was also needed for producing dyes because coal-tar dye production was a virtual German monopoly at the outbreak of hostilities. By 1930, less than four decades after the first byproduct coke oven was installed, more than 12,000 byproduct ovens were operated in the United States. They accounted for about 95% of all coke produced in the United States that year.

Description. A byproduct coke oven is a chamber about 20 inches wide, 10–20 feet high, and 35–50 feet long. (The ovens are so narrow because the coal charge is a poor conductor of heat, and wider ovens would be difficult to heat through to the center.) A group of these ovens is arranged together in a *battery.* Depending on the production needed from a particular coke plant, a battery may have anywhere from 15 to more than 100 ovens. A typical charge of coal ranges from 10 to 35 tons and is determined by the overall size of the oven. The heat needed for the carbonization process is obtained by burning gas in flue chambers that are located in the walls between adjacent chambers.

The coal is crushed so that about 80% will pass through a screen having ⅛-inch openings, and the maximum size of any particle is 1 inch. The crushed coal or coal blend is loaded into a larry car that runs along the top of the battery and feeds the coal into charging holes in the tops of the ovens.

When the coal drops into the oven, the particles lying against the heated walls begin to soften, decompose, and resolidify to coke. The heat from the hot walls gradually penetrates through the mass of coal. This heat transfer from the walls establishes three zones in the oven:

1. a layer of coke closest to the walls,
2. a layer of plastic coal undergoing carbonization, and
3. an innermost layer of coal that has not yet reached carbonization temperature and remains for the moment unreacted.

As time passes, the plastic layers move in from each wall toward the center of the oven and eventually merge. The coalescence of the plastic masses and the solidification to coke cause a crack to form in the solid mass in the middle of the oven. Thus, the largest coke pieces from a byproduct oven are one-half of the width of the oven. The coking process is judged to be complete when the temperature at the center is about equal to that at the walls. The coking of a single charge of coal takes 15–20 hours.

When coking is finished, the product is removed from the oven by a mechanical pusher. The oven width tapers slightly to make the pushing operation easier. The hot coke moves to a quenching station where it is cooled by a water spray. It is then moved to a wharf from which it can be taken for use.

Advantages. Because no air is admitted to the byproduct oven, none of the volatile carbonization products are lost by burning and instead can be collected and recovered as valuable byproducts. The byproduct oven also yields more coke than the beehive oven. It converts up to 75% of the coal charged, whereas the beehive oven converts about 66%. Although this difference may seem minor, the byproduct oven has an appreciable advantage when the enormous tonnage of coal converted to coke each year is considered. For every 10 million tons of coal, the byproduct oven produces 900,000 more tons of coke than the beehive oven produces.

Coking Temperatures. Coking takes place at temperatures around 2000 degrees Fahrenheit. At this temperature, all of the byproducts can be piped out of the oven as gases to be condensed and separated. The hot gases coming out of the oven pass through a spray of water that drops their temperature from about 1500 degrees Fahrenheit to less than 200 degrees Fahrenheit. The immediate effect of this temperature drop is the condensation of a black, viscous, gooey substance known as *coal tar.* Although the tar might seem noxious at first, further chemical treatment of coal tar has produced more than 5000 dyes, medicines, flavorings, and perfumes.

Byproduct Recovery. A block flow diagram of the recovery of byproducts from a coke oven is shown in Figure 8.2. Once the tar has been removed from the gas, treating the gas with sulfuric acid or phosphoric acid recovers ammonia. The ammonium sulfate or ammonium phosphate thus produced can be used as fertilizer. Ammonia can also be converted to nitric acid, which is the starting material for the production of many explosives. Ammonia itself can be used in chemical processing, for example, in the manufacture of synthetic fibers.

Further cooling and treatment of the gas produces a material called *light oil,* which is a mixture of benzene, toluene, and xylene as well as a few other compounds such as naphthalene. Benzene is a useful solvent and can be converted to polystyrene, polystyrene foam, synthetic rubber, and aniline, which itself is the parent of a host of compounds. Toluene is also used as a solvent and is the precursor of diverse substances such as saccharin and 2,4,6-trinitrotoluene (TNT). Naphthalene is used in mothballs.

An aqueous liquor condenses from the gas stream. This liquor is a solution of ammonium carbonate, ammonium sulfides, phenols, cresols, and cyanides. It is treated either to recover these compounds or to reduce them to levels acceptable for discharge into the environment.

The gas, principally a mixture of hydrogen and methane, has about one-half of the heating value of natural gas. In the days when cities had large distribution networks for manufactured gas, the coke-oven gas could be sold and blended into distribution systems for home heating or illumination. Today, most of the coke-oven gas is used on site as fuel either in the flue chambers in the coke-oven batteries or in the furnaces used for heat treating finished steel.

A modern byproduct coke oven yields about 1500 pounds of coke, 11,000 cubic feet of gas, 8–10 gallons of tar, 3–4 gallons of light oil, and 25 pounds of chemicals (mostly ammonium compounds) per ton of coal.

How Coke Is Used

Blast Furnaces

Description. About 90% of all the coke produced is used in blast furnaces for smelting iron ore. A *blast furnace* is essentially a shaft 60–100

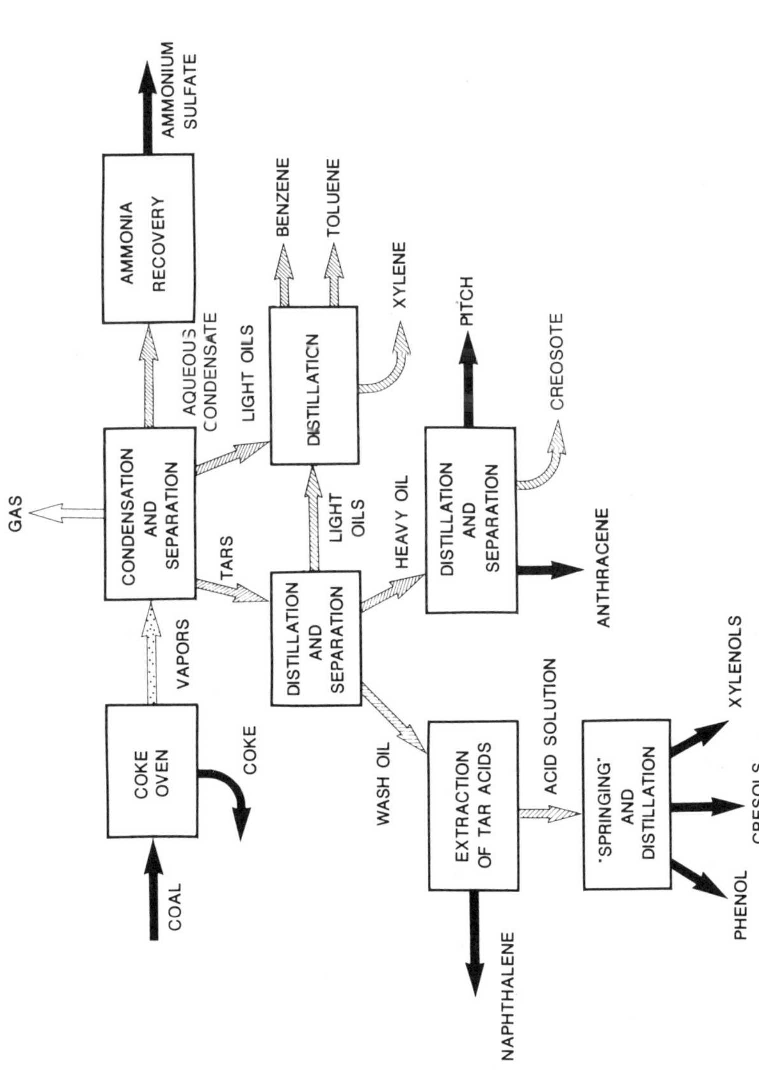

Figure 8.2. Volatile byproducts from a coke oven are treated with a sequence of relatively simple operations to separate about 12 important materials. The products of the operations shown here can be used to make hundreds of other useful chemicals.

feet high formed from a steel shell lined with refractory bricks. The furnace has provisions for charging a mixture of coke, iron ore, and a flux (usually limestone) in the top; for blowing a blast of hot air into the bottom; and for removing the molten iron and slag. The ingredients needed to produce 1 ton of iron are about 1.8 tons of ore, 1200 pounds of coke, and 500 pounds of limestone.

Each coke, ore, and flux charge gradually settles through the furnace as the smelting process consumes the preceding charge. The coke burns in the region of the air blast. The overall reaction is the combustion of the carbon in the coke by oxygen to form carbon monoxide. Some of the carbon may be burned to carbon dioxide, but this gas will react almost immediately with the hot carbon nearby to make carbon monoxide. The combustion of the coke serves two purposes:

1. The carbon monoxide is the active reagent responsible for the reduction of the iron oxides to metallic iron.
2. The heat liberated melts the iron and the slag.

Enough excess coke is charged to the furnace so that appreciable carbon monoxide exits the furnace along with other gases. Dust or particulate matter is removed from this gas stream. The clean gas is burned to preheat the air blast to about 1200 degrees Fahrenheit.

Pig-Iron Production. The product of the blast furnace is *pig iron.* This name derives from the once-standard practice of pouring the molten metal into a system of molds that looked like a sow nursing a litter of piglets. Pig iron has a carbon content of 4.5%, which is high for a metal, and also contains some silicon, manganese, phosphorus, and sulfur. It is very brittle and virtually useless by itself. Most pig iron is converted to steel by reducing the level of impurities and sometimes adding alloying agents such as nickel or chromium. Pig iron is also used in foundries to make cast iron.

Properties of Blast-Furnace Coke. Exact specifications for the ideal blast-furnace coke do not exist because of the various designs and operating conditions of blast furnaces and variations in the composition of the charge. The most important properties are porosity, strength, size, and carbon content. The absence of undesirable properties and the consistency of quality are important as well.

Porosity. The escape of the volatile byproducts through the plastic mass of carbonizing coal creates a porous, cellular structure in the resulting coke. The porosity of the coke allows it to be readily permeated by the air blast; this condition results in rapid combustion. The rapid burning of the coke in the lower regions of the blast furnace is desirable for establishing a high temperature in the bottom of the furnace for melting the iron and slag.

Strength. A high mechanical strength is even more important than porosity. Coke receives rough handling during transportation and charging into the furnace. Then, as it descends the furnace, it is exposed to abrasive action by the walls and by other materials in the charge. The coke pieces must retain most of their size to maintain an open structure in the charge so that the air blast and combustion gases can pass through the furnace. The coke must also be strong enough to resist crushing. As it approaches the bottom of the furnace, the coke must support many tons of charge mixture lying on top of it. The ability of the coke to support the charge above it is also important for maintaining an open structure throughout the charge. Bituminous coal cannot withstand the crushing forces experienced by the coke and will collapse to a packed powder impermeable to the air blast. In the worst case, the loss of bed permeability will shut down the furnace.

Size. The need to maintain a good flow of gases through the charge also means that the size of the coke pieces must be consistent. If a mixture of small and large pieces is used, the smaller pieces or fine particles tend to fill the void spaces in the charge and impede the flow of gases. This situation can cause a number of problems in the furnace, including a lower melting rate, an increase in carbon pickup by the metal, and higher coke consumption.

Carbon Content. Coke generally contains 85–90% carbon. A high carbon or fixed carbon content is desirable because in general it is associated with a high heating value. As the heating value of coke increases, the amount needed to achieve a desired heat output decreases.

Undesirable Qualities. Undesirable qualities of coke are a high ash content and high sulfur and phosphorus contents. The ash content increases the volume of slag to be handled. An increase in ash content of coke tends to reduce the temperature of the metal. Sulfur and phosphorus in coke enter the iron and must later be removed when the iron is converted to steel.

Consistency of Quality. Consistency from day to day is an important criterion of coke quality. Little variation in the coke means that the furnace operators can be confident of reproducing the temperatures in the furnace and the quality of the pig iron for a given set of operating conditions. Receiving coke of extra good quality can be almost as disastrous as receiving coke of poor quality.

Steel Industry

The coke industry is inextricably linked with the steel industry: If steelworkers strike, mines producing coking coal may have to close because of a lack of demand. If miners strike, especially for a long time, loss of supplies of domestic coking coal may force the steel industry to look to imports or to curtail production.

Changes in technology may alter the relationship between coal mining and steel making. For years, the rule of thumb was that 1 ton of coke was needed to make 1 ton of pig iron. However, this so-called *coke rate* has dropped significantly. From 1950 to 1970, innovations in blast-furnace practice reduced the coke rate from 1 ton of coke to about 1250 pounds of coke per ton of pig iron. Thus, these innovations decreased the demand for coke and coking coal. On the other hand, changing technologies can increase the demand for coke. For example, the open-hearth furnace used in steel making for a century is rapidly being replaced by the basic oxygen process. The open-hearth furnace uses about equal quantities of scrap steel and pig iron. A typical open-hearth charge is 48% scrap steel and 45% pig iron, and the rest is small quantities of limestone, iron ore, and spiegeleisen, an additive containing manganese. However, the basic oxygen furnace charge is 30% scrap steel and 70% pig iron. Thus, these furnaces have increased the demand for coke because they require more pig iron than the open-hearth furnaces.

Today, the steel industry is in a transitional period. To some observers, steel is a dying smokestack industry of the U.S. rust belt. (The *rust belt* is the region of the United States, mainly in the Northeast, where heavy industry, for example, steel, automobiles, machine tools, and railroads, used to be very prosperous and is now in decline.) To others, recent innovations such as continuous casting processes and minimills suggest that steel can emerge from this transitional period as a healthy but radically changed industry. The outcome is likely to have a lasting effect on the bituminous coal fields.

Foundries

About 90% of the coke produced is used in blast furnaces for making pig iron, and the rest is used in several ways. One of the more important of these other uses is as a fuel in foundries.

Cupolas. Iron is melted for casting in furnaces called *cupolas.* A cupola is a small version of a blast furnace; the main difference is that no smelting is done in a cupola. Instead, the cupola is charged with a mixture of coke, pig iron, scrap metal, and sometimes alloying agents. Heat from burning the coke melts the charge and produces various grades of cast iron.

Properties of Foundry Coke. Many of the specifications of foundry coke are similar to those of blast-furnace coke: low ash, low sulfur, high fixed carbon, and good mechanical strength. Foundry coke should have a much lower porosity than blast-furnace coke. In the blast furnace, the coke must have a high reactivity to make carbon monoxide, which reacts with the iron ore to liberate the iron. In the cupola, carbon monoxide is not needed because no ore reduction occurs. Burning the carbon to carbon dioxide liberates about 3.5 times more heat per pound of carbon than burning the carbon to carbon monoxide. The real need in the cupola is for heat. Carbon dioxide is produced when the coke surface is burned. If the carbon dioxide were forced through the pores of the coke, most likely it would be converted to carbon monoxide by reaction with the hot carbon forming the walls of the pores. The less reactive, lower porosity coke is less likely to ignite in the upper regions of the cupola, where the heat from its burning would be lost. Using blast-furnace coke in a cupola would result in the waste of carbon monoxide in the gases leaving the cupola, require a higher coke rate, and not get the metal as hot.

Coke Breeze. Coke pieces smaller than one-half to three-fourths of an inch are known as *coke breeze.* Because the only difference is size, coke breeze has the same resistance to rough handling, high heating value, and low sulfur content as coke; these properties make it an excellent fuel. Coke breeze is used to sinter iron ore into pellets for feeding to blast furnaces and is used as a boiler fuel for generating steam. These same qualities of coke and coke breeze are very desirable in fuels used for domestic heating. In fact, before World War II, about 20–25% of the coke produced in the United States was used as a household fuel. The growth of the interstate pipeline network and consequent availability of cheap natural gas after World War II virtually wiped out this market for coke.

Reaction of Coke with Air and Steam. Coke can react with air and steam to produce a mixture of hydrogen and carbon monoxide called *water gas.* Burning coke in a limited supply of air forms *producer gas,* in which the main combustible component is carbon monoxide. Water gas and producer gas were once used extensively as fuels; however, cheap natural gas virtually destroyed their manufacture.

Reaction of Coke with Calcium Oxide. When calcium oxide reacts with coke at the extreme temperatures of an electric arc furnace, calcium carbide is produced. Calcium carbide reacts with water to form acetylene. Small quantities of calcium carbide are used in carbide lanterns, which mix the solid with water and burn the acetylene. Calcium carbide is occasionally used for repelling—or asphyxiating—moles by spooning the compound into their tunnels and is commonly used to fuel welding torches. On a larger and much more important scale, acetylene was once used as a starting material for preparing many important chemicals, including ethyl alcohol and acetic acid, as well as various end products, such as plastic and synthetic rubber. In Germany during World War II, when raw materials from conventional sources were becoming increasingly scarce, the monthly production of calcium carbide was more than 100,000 tons. This calcium carbide was used to make acetylene, which was then used to make other chemicals and end products. The ultimate sources of these acetylene-derived products were three readily available, inexpensive materials: limestone, coal, and water.

Low-Temperature Carbonization

Coking is the most important example of high-temperature carbonization processes, that is, carbonization processes occurring at temperatures up to 1650–2000 degrees Fahrenheit. Low-temperature carbonization processes have a maximum temperature usually in the range from 850 to 1350 degrees Fahrenheit. Low-temperature carbonization is used mostly to produce smokeless fuel for domestic use or small industrial installations. In the past, low-temperature carbonization was used to produce tar for the chemical industry. It was also used extensively to produce gas for town lighting, but the availability of inexpensive natural gas has significantly reduced interest in this application.

Char

Low-temperature carbonization processes are generally designed to use lignite, subbituminous coal, or high-volatile bituminous coal. The solid product of low-temperature carbonization is *char* or *semicoke,* which is reactive and easily ignited. In addition to being used as a smokeless fuel, char can be used in gasification processes, blended with coals to make a coke-oven feed, or used as power-plant fuel. Low-temperature carbonization char is not suitable for use in blast furnaces or cupolas; however, it is occasionally used in metallurgical processes such as the production of ferrosilicon, an alloying agent used to make corrosion-resistant cast iron.

Tar

As coal is heated, the volatile carbonization products start to evolve around 750 degrees Fahrenheit. In a low-temperature carbonization process, the walls of the *retort* or oven used for carbonizing may not be much hotter than this temperature. The volatile products tend to pass out of the oven without much additional reaction. During high-temperature carbonization, the products originally evolved from the coal come into contact with much hotter surroundings and decompose or react further. Chemically, the low- and high-temperature tars are very different: Low-temperature tars contain predominantly aliphatic compounds, whereas high-temperature tars contain mainly aromatic compounds. The type of tar that is preferable is determined by the anticipated use. High-temperature tars are desirable for plants designed to recover byproduct chemicals, such as from coke-oven tar. On the other hand, low-temperature tars can be refined into lubricants, heavy heating oils, and even gasoline. The yield of tar is greater from low-temperature carbonization because less decomposition occurs as the tar vapors pass out of the oven. A low-temperature carbonization process can yield 70–80% char and 7–10% tar.

Gas

Another distinction between low- and high-temperature carbonization is the amount and type of gas produced. The gas yield in low-temperature carbonization is only about 25% of the yield from high-temperature ovens; however, the gas is richer in methane and lower in hydrogen and thus has a higher heating value.

Coal-Smoke Fog

Using low-temperature carbonization to make smokeless fuel for use in homes and small commercial establishments is particularly important in areas having a high population density and relying on coal as a fuel, particularly coals that have high contents of volatile matter. As coal is heated to its kindling temperature in a furnace or fireplace, it carbonizes. Some products of carbonization, such as benzene and naphthalene, undergo only partial combustion in home hearths or furnaces. The unburned carbon escapes as smoke or soot. Once in the atmosphere, the particles of soot can adsorb sulfur dioxide, which is eventually transformed to sulfuric acid. The acid particles then readily absorb moisture in humid air and form a fog. The evaporation of the moisture droplets in the fog is retarded by the acid and by thin films of oils or tars that also escaped combustion. Coal smoke not only promotes and initiates the formation of a fog, it also makes the fog thicker and harder to dissipate.

The fogs of London and other heavily industrialized areas of Great Britain are the prime examples of coal-smoke fogs. Today, these fogs have a romantic or nostalgic patina by association with the exploits of Sherlock Holmes and Dr. Watson being pursued by Moriarty through fog-shrouded London. The reality, however, was much grimmer. Charles Dickens describes the reality on the opening page of *Bleak House*[2]:

Fog everywhere. Fog up the river, where it flows among green aits and meadows; fog down the river, where it rolls defiled among the tiers of shipping and the waterside pollutions of a great (and dirty) city. Fog on the Essex marshes, fog on the Kentish heights. Fog creeping into the cabooses of collier-brigs; fog lying out on the yards and hovering in the rigging of great ships; fog drooping on the gunwales of barges and small boats. Fog in the eyes and throats of ancient Greenwich pensioners, wheezing by the firesides of their wards; fog in the stem and bowl of the afternoon pipe of the wrathful skipper, down in his close cabin; fog cruelly pinching the toes and fingers of his shivering little 'prentice boy on deck.

Smokeless Fuel

The fogs just described were the impetus for producing a smokeless fuel. Because volatile matter is the component of coal that does not burn completely, the objective of a process for making smokeless fuel is to reduce the volatile matter content without severely reducing the reactivity of the solid. The temperatures of volatile matter evolution depend somewhat on the rank of the coal and the exact conditions of heating. The evolution of volatile matter from bituminous coal usually begins at about 750 degrees Fahrenheit and ends at 1100 degrees Fahrenheit. Heating the char or semicoke higher than 1500 degrees Fahrenheit tends to reduce its reactivity so that it does not burn well in small units such as domestic furnaces. Lignites, subbituminous coals, and high-volatile bituminous coals are preferred for low-temperature carbonization for two reasons:

1. The char from these coals tends to be almost as reactive as the parent coals themselves.
2. Coking coals, that is, medium- or low-volatile bituminous coals, swell and stick to or clog the carbonization equipment.

Commercial Processes

Over the years, a wide variety of processes have been developed for low-temperature carbonization. Two commercial processes are used in Great Britain for producing smokeless fuels by low-temperature carbonization.

In the Homefire process, high-volatile bituminous coal is crushed to ¼-inch particles and devolatilized for 20 minutes at 800 degrees Fahrenheit in a fluid-bed reactor. This relatively quick carbonization reduces the volatile matter content of the coal to about 20%. The hot char is fed directly to a hydraulic press, where it is formed into briquettes that are a premium-grade domestic fuel.

The Phurnacite process starts with fines of low-volatile bituminous coal blended with pitch and formed into briquettes. The briquettes are then carbonized for about 4 hours at 1400 degrees Fahrenheit. The tars and gases are recovered, similar to their recovery in a coke oven. The solid product is strong, hard briquettes that make excellent domestic furnace fuel.

Notes

[1] Kipling, R. In *The Macmillan Book of Proverbs, Maxims and Famous Phrases* Stevenson, B., Ed.; Macmillan: New York, 1948.

[2] Dickens, C. *Bleak House*; Signet: New York, 1964.

Chapter 9

Chemicals from Coal

Early Uses of Coal Chemicals

The carbonization of coal produces three substances:

- a solid, namely, coke;
- a flammable gas; and
- a black, sticky, foul-smelling liquid.

The liquid is called *coal tar.* Coke, as discussed earlier, was a key ingredient in the Industrial Revolution of the 18th century; it replaced the increasingly scarce charcoal as the fuel for iron smelting. In 1792, William Murdock, a Scottish engineer who worked for James Watt, experimented with lighting his home with the gas made from coal carbonization. He was so successful that in 1798, the Boulton and Watt factory in Birmingham was lighted with gas. Frederick Winsor, a German entrepreneur, obtained a British patent for gas manufacturing in 1804; in 1807, his firm introduced gas lights for public streets, starting with the Pall Mall in London. By 1816, virtually all of London had gas lights. Gas-lighting soon spread throughout Europe and the United States.

The early decades of the 19th century witnessed a growing demand for two of the three carbonization products: gas and coke. The other carbonization product, coal tar, was becoming a nuisance. Few uses existed for coal-tar products. Distillation of the tar produced naphtha, which is a good solvent for organic materials, for example, the rubber used by Charles Mackintosh to waterproof coats. The thick pitch remaining after distillation was used for tarring roads. Coal tar was also used as a wood preservative;

one application was the treatment of railroad ties. These pioneering uses of coal tar consumed only a portion of the total production. The rest was discarded as conveniently as possible, which in those days often meant dumping it into the nearest stream. The unused coal tar was a major contributor to industrial pollution in the early 1800s.

Establishment of Research Institutions

Around the middle of the 19th century, chemists began probing to determine the components of coal tar. Tar was subjected to distillations and extractions and was gradually refined into somewhat pure compounds that could be tentatively identified. This early work clearly showed that coal tar was a complex mixture of dozens of components.

Before the mid-19th century, the foremost chemical laboratory in the world was at the University of Giessen in West Germany under the direction of Justus von Liebig. Students went to Giessen from Great Britain, the United States, and many European countries. Many of Liebig's graduates had careers of great prominence in chemistry; their names became familiar terms in chemistry: Erlenmeyer, Hofmann, Kekulé, Wurtz, and Regnault. Among this group was a student who had entered Giessen to study law and languages but who had become attracted to chemistry instead; his name was August von Hofmann.

While Hofmann was a student in Liebig's laboratory, he began research on the chemistry of coal tar. One of Hofmann's projects was to determine whether or not a compound called kyanol, isolated from coal tar in 1834 by Friedlieb F. Runge, was the same compound as the aniline obtained by degrading the dye indigo. Hofmann showed that the two compounds were the same and also that they were identical to the aniline produced by nitrating and then reducing benzene, another constituent of coal tar.

The renown of Liebig and his school inspired others to establish laboratories for education in chemistry. In England, Prince Albert was instrumental in founding a new institution, the Royal College of Chemistry. Liebig was well-known in Great Britain and was a native of the same region of Germany as Prince Albert. On Liebig's advice, August von Hofmann became the first director of the Royal College of Chemistry. The position allowed Hofmann to continue his work on coal tar because it was in plentiful supply in Great Britain.

Hofmann patterned the activities of the Royal College after the successful model of Giessen. Unfortunately, Hofmann ran afoul of the interests of the financial supporters of the Royal College. They could see little reason for Hofmann's strong commitment to fundamental research; instead, they urged an emphasis on research applied to problems directly relevant to established industrial practice. As events soon proved, this attitude was disastrously shortsighted.

Research in organic chemistry during the mid-19th century focused on the isolation of compounds from natural sources and the development of methods to synthesize the compounds. Runge, for example, had a prolific career separating components of coal tar. In 1820, before his work with coal tar, Runge was the first chemist to isolate caffeine from coffee. One of Hofmann's main interests was the chemistry of organic nitrogen compounds, which occur extensively in plant material.

Synthesis of Quinine

Since the mid-1600s, the bark of the cinchona tree has been used to treat malaria. Around 1820, two French chemists, Pierre Pelletier and Joseph Caventou, isolated quinine, which is one of the active ingredients of cinchona bark. Hofmann asked his 18-year-old laboratory assistant, William Henry Perkin, to develop a synthesis for quinine.

In those days, developing a synthesis for an organic compound was truly a voyage into uncharted waters. Today, the strategy for an organic synthesis can be based on a knowledge of the three dimensional molecular structures of the desired product and available starting materials, as well as a knowledge of the kinds of organic reactions that transform one structure into another. The molecular structures derive from the concept of the aliphatic tetrahedral carbon atom, developed independently by Jacobus van't Hoff and Joseph Le Bel in 1874, and the concept of the aromatic benzene ring, proposed in 1866 by Friedrich Kekulé. Before organic chemists had any idea of the structural complexity of organic molecules, one approach to planning an organic synthesis was the addition and subtraction method. If chemists knew the molecular formulas of the materials they wanted to make and of their starting materials, they could, according to the principle of addition and subtraction, figure out what other materials they needed to add to the reactants or subtract from the products to arrive at the desired result.

Perkin knew that the molecular formula of quinine was $C_{20}H_{24}N_2O_2$. It seemed, by the addition and subtraction method, that the synthesis of quinine should be a straightforward oxidation of the chemical allyltoluidine ($C_{10}H_{13}N$). Two molecules of allyltoluidine should furnish the 20 carbon atoms, 24 hydrogen atoms, and 2 nitrogen atoms needed for one molecule of quinine. By reacting allyltoluidine with a source of oxygen, the two oxygen atoms needed for the quinine molecule could be obtained, and the two "extra" hydrogen atoms (the two allyltoluidine molecules supply 26 hydrogens, but the quinine molecule incorporates only 24 hydrogens) could be used to form a molecule of water, H_2O. Perkin attempted the synthesis during his Easter vacation in 1856 in a laboratory he set up in his home. Because the structural concepts of Kekulé, van't Hoff, and Le Bel were not yet known, Perkin was undeterred by the fact that the molecular formula

$C_{20}H_{24}N_2O_2$ could easily represent dozens of different structures, only one of which would be quinine.

When Perkin attempted his synthesis, the product turned out to be an intractable, brownish black mess. To achieve a better understanding of what was happening in the reaction, he employed a strategy that is still a favorite today: He repeated the reaction with a material of simpler composition to try to eliminate any unwanted reactions that might lead to byproducts. Perkin chose to repeat the reaction with aniline. Once again, the product was a dark, gooey mess. When Perkin started to clean up his equipment with alcohol, he found that the alcohol became purple. Perkin saved the alcohol, which he evaporated to recover purplish black crystals. These crystals proved to be a fairly good dye, generally called mauve and sometimes referred to as mauvine or aniline purple.

Dyes from Coal Tar

The serendipitous preparation of mauve while attempting to synthesize quinine was the start of an incredible growth of research and development in organic chemistry, based largely on a foundation of coal tar. Perkin quit school against the advice of his mentor Hofmann; with the financial support of his father and brother, he began to manufacture dyes.

Because of the efforts of the Perkins, Great Britain developed a commanding position in the manufacture of dyes from coal-tar chemicals. The British dye industry seemed to have an invincible position in world trade, just as the British navy had an invincible position on the high seas. But then the Germans made a crucial strategic decision to invest heavily in fundamental research on the chemistry of coal and coal tar. The British, meanwhile, continued to rely extensively on empirical methods, the very philosophy that the supporters of the Royal College were urging Hofmann to use. In addition to the decision to support basic research in coal chemistry, the Germans also realized the importance of attracting the best possible scientific talent. In 1864, Hofmann went home to Germany, first to a magnificent home and laboratory provided for him in Bonn, and then a year later to even grander accommodations in Berlin.

To the modern business manager or government planner, the results of the German decision may seem to have taken a long time to develop. Nevertheless, they did come, relentlessly and implacably. By 1875, the German dye industry was at least the equal of the British dye industry. Within another 10 years, Germany was supreme in the world. At the outbreak of World War I, the scarlet "redcoats" of the British army and the red trousers of the French army were colored with German dyes. Coal tar was the basis of a thriving chemical industry for two reasons:

1. Coal tar is a plentiful byproduct of two other industries, namely, coking and gas production.
2. Coal tar contains a great variety of chemicals that are useful in organic syntheses and that can be obtained from the tar by straightforward procedures.

The refining of coal tar involves mostly distillation and treatment with sodium hydroxide and sulfuric acid, which are inexpensive.

Separation of Coal Tar

The composition and properties of coal tar depend on the carbonization conditions that were used and on the specific coal that was carbonized. The processes used for separating the tar depend on the composition of the tar and on the specific products that are desired. A typical separation might be as follows: The tar is distilled to make volatile oils. The light oil, which is the most volatile, contains benzene, toluene, and xylenes. These compounds are useful as solvents and serve as the starting point for the synthesis of a host of other materials. The higher boiling oil is treated with a solution of sodium hydroxide to extract tar acids. The tar acids are mainly hydroxyl derivatives of benzene or naphthalene: phenol, cresols, xylenols, and naphthols. These compounds dissolve into the sodium hydroxide and are then removed from the solution by treating it with sulfuric acid or carbon dioxide. The mixture of tar acids is separated into its individual components by careful distillation. Treating the oil with a sulfuric acid solution extracts a family of tar bases. These are mostly compounds in which a nitrogen atom is attached to the benzene or naphthalene ring, for example, aniline, or is a part of the ring, for example, pyridine. The tar bases are removed by treating the sulfuric acid solution with lime or sodium hydroxide; they are then separated from each other by distillation. The washed oil, left after removal of tar acids and bases, is distilled to naphtha and naphthalene. Continuing the distillation of the tar produces creosote oil; a mixture of phenols, naphthalene, and anthracene used as a wood preservative; and anthracene oil, which contains mostly anthracene and phenanthrene and is the starting material for a family of dyes. The material remaining after distilling the anthracene oil is *pitch*.

Pitch

Up to one-half of the total amount of tar can remain as pitch. The properties of the pitch are determined by the extent of distillation, which is governed by the expected use of the pitch. For a relatively soft pitch, which might be used for road tar or roofing tar, the distillation is stopped at an early point so that some relatively fluid compounds are left in the pitch. Proceeding further produces a thicker material that can be used as a binder in making fuel briquettes or carbon electrodes in electric metallurgical furnaces. Extensive distillation of the tar results in a very hard material used as fuel, or a *pitch coke.*

Primaries and Intermediates

Hundreds of compounds can be isolated from coal tar by careful separation and analysis in the laboratory. The 1945 edition of *Chemistry of Coal Utilization* lists 345 such compounds. For industrial purposes, the 10 most important chemicals from coal tar are called *crudes* or *primaries.* These include benzene, naphthalene, anthracene, toluene, and phenol. The primaries are the versatile building blocks used by the synthetic organic chemist; they collectively yield more than 300 so-called *intermediates.* The intermediates have been used to make more than 900 individual dyes, which at one time were on the market under almost 6000 brand names. Dye chemistry was the impetus for developing the coal-tar industry. The dyeing of fabrics was one of the earliest chemical techniques developed by prehistoric people. Until Perkin's discovery of mauve, virtually all dyes were made by extracting colored materials from plants or animals. Dyes from natural sources sometimes do not hold their color very well (they are not *fast*) and can produce different results when applied to different textiles. In addition, some natural dyes are expensive.

Synthetic Dyes

The first synthetic dye was picric acid, which was prepared in 1771 by oxidizing indigo. The first synthetic dye to be prepared from coal tar was aurin, a yellow dye made by Runge in 1834. In 1855, a method was developed to make picric acid by nitrating phenol from coal tar. Neither picric acid nor aurin achieved much commercial use as a dye. In fact, picric acid became far more widely used as a powerful explosive. Mauve was the first commercially successful synthetic dye.

Some of the early synthetic dyes were not much superior to the natural dyes until greater colorfastness and color range were developed. The textile industry showed reluctance to use dyes that were not from natural sources. As research continued on the development of coal-tar dyes, colorfast dyes were produced. These dyes can be selected for use on a particular fabric and are cheaper and easier to obtain than natural dyes. The coal-tar research laboratories have produced about 900 colors. Some examples are rosaniline, Turkey red, vital red, azo scarlet, imperial blue, quinoline blue, Victoria blue, indigo, mauve, fuchsin, Tyrian purple, Bismarck brown, and Hofmann violet.

Alizarin

In 1828, chemists in France isolated the active ingredient in the dye prepared from madder roots; the ingredient was a compound called *alizarin.* About 40 years later, studies of the molecular structure of alizarin showed that the basic framework was the anthracene ring system. Anthracene is one of the coal-tar primaries. The knowledge that the structure of alizarin was derived from anthracene led to competition among coal-tar chemists in Great Britain and Germany to be the first to synthesize alizarin. The matter was decided by a single day. The German chemists Karl Graebe and Carl Liebermann applied for a patent on June 25, 1869, one day before Perkin.

The history of synthetic alizarin is typical of the commercialization of research findings. The determination of the structure of alizarin took decades. Then the synthesis of alizarin from anthracene had to be developed in the laboratory. Once a successful synthetic route was found, a commercial process was developed, in this case in about 6 months. About 5 years after the development of the commercial process, annual sales of synthetic alizarin reached $8 million. The commercialization of a synthetic alizarin process completely devastated the commercial farming of madder, which accounted for some 300,000 acres in Europe and the Near East. This farm land became available for other crops, including foodstuffs, but the short-term effect was catastrophic until the transition from madder farming to other crops was completed.

British and German Industries

Perkin abandoned the field to the Germans in 1874. Part of his frustration stemmed from the sad state of chemical education in Great Britain. From 1865 to 1874, none of the universities in England had a professor of organic chemistry. Hofmann, meanwhile, had the opportunity to recruit many well-trained young scientists with doctorates from German universities. The difference in viewpoint, that is, a reliance on empirical methods in Great Britain versus fundamental research in Germany, also resulted in radically different employment opportunities for chemists. Shortly after the beginning of the 20th century, six factories in England were making synthetic dyes. Collectively, they employed 35 chemists. At the same time, a German company, the Hochster Farbwerke, employed more than 300 chemists.

Indigo

Another case of a synthetic dye replacing a natural dye was indigo. The synthesis of indigo required 17 years of research and was supported by the Badische Anilin und Soda-Fabrik at a cost of $5 million. The active agent in natural indigo was identified as *indigo carmine.* Further work showed that the structure of indigo carmine was related to another coal-tar primary, namely, naphthalene. The synthesis of indigo carmine required that the naphthalene be oxidized to phthalic acid. Chemists knew how to make phthalic acid from naphthalene, but the reaction took long and yielded little phthalic acid; a commercial process based on this reaction would not have been profitable. Then, during a test of the phthalic acid synthesis, a thermometer broke. The mercury from the broken thermometer proved to be an excellent catalyst for the naphthalene oxidation. This finding opened the road to commercial production of synthetic indigo and at the same time ruined a million acres of indigo plantations in India.

The development of the process for synthesizing indigo took a long time, cost a lot of money, and required a lot of faith and patience. For such an investment, Badische obtained a 96% share of the world indigo market. The cost of indigo plummeted more than 25-fold, from $4 to 15 cents per pound. When the synthetic indigo process was developed in 1897, Germany was paying $3 million per year to import indigo. By 1914, Germany was exporting indigo to the rest of the world and achieved annual sales of about $13 million. The research on indigo made a $16-million turnabout in Germany's balance of payments.

Other Compounds from Coal Tar

Toluene

The manufacture of chemicals from coal tar is not limited to synthetic dyes. Each of the coal-tar primaries can be used for making a wide range of products. Toluene, for example, is useful as a lacquer solvent in the paint industry and as a solvent to extract waxes and other substances from plants. Used as the starting material in chemical syntheses, toluene can be converted into an explosive, trinitrotoluene (TNT); a preservative, benzoic acid; an antiseptic, chloramine-T; a perfume and flavoring agent, benzaldehyde (artificial oil of almond); and a nonnutritive sweetener, saccharin. Syntheses of several classes of compounds, including explosives, medicines, flavors, and perfumes, were extensions of the chemistry used in dye research.

Nitro Compounds

When aromatic compounds such as those found in coal tar react with nitric acid under appropriate conditions, the products are nitro compounds. The gaseous diatomic nitrogen molecule is very stable; thus, when compounds containing nitrogen atoms decompose, gaseous nitrogen is often one of the products. If the decomposition occurs rapidly and heat is evolved, a large volume of gaseous products forms rapidly. If these products are temporarily confined, such as in an artillery shell, a great increase in pressure occurs. The surrounding material shatters to release the pressure, and we observe this event as an explosion. TNT is a well-known explosive made by nitrating toluene. Nitramine is a nitro derivative of aniline. The nitration of phenol produces picric acid. Picric acid is a chemical that can have both beneficial and harmful effects, depending on how people decide to use it. During World War I, picric acid was a common ingredient of highly explosive shells that produced a large cloud of black smoke when they burst. These shells were sometimes called Jack Johnsons in reference to the powerful punches of the Negro heavyweight boxing champion. Picric acid was also widely used as an antiseptic for treating combat wounds.

Medicinal Products

Many medicinal products derive from coal tar. Phenol has been used as an antiseptic; it is better known in this application by its common name, carbolic acid. Cresols are used in antiseptic or disinfectant soaps and washing materials. Coal-tar derivatives have been used in the synthesis of antibacterial compounds, for example, sulfanilamide and chloramine-T; drugs used for the relief of mild pain or fever, for example, aspirin, phenacetin, acetanilide, and antipyrine; and local anesthetics, for example, procaine hydrochloride and amylocaine hydrochloride. The picolines, derivatives of coal-tar pyridine, are used to prepare nicotinic acid, which is an antipellagra vitamin, and Isoteben, which is a drug used to treat tuberculosis.

Flavorings and Perfumes

Coal tar has a characteristic phenolic odor; disinfectants such as Creolin (prepared from refined coal-tar oils) provide mild examples of this smell. The black viscous coal tar has been the source of flavorings and perfumes, including the artificial flavors of vanilla, almond, lemon, and wintergreen. The diversity of substances that can be isolated from coal tar or made from its derivatives, as well as the perceived role of coal tar as one of the founts of the emerging age of scientific wonders, once even made the pages of *Punch*:

> *There's hardly a thing that a man can name*
> *Of use or beauty in life's small game,*
> *But you can extract in alembic or jar*
> *From the "physical basis" of black coal tar—*
> *Oil and ointment, and wax and wine*
> *And the lovely colors called aniline,*
> *You can make anything from a salve to a star,*
> *If you only know how, from black coal tar.*

Expansion of Research

The chemistry of coal tar was the stimulus for a great expansion of research in organic chemistry during the late 19th century and early 20th century. The prospect of preparing useful substances from what was once considered a noxious byproduct attracted not only scientific talent but also capital for the support of research and the development of industrial processes. The chemist no longer was merely a person who analyzed substances to find out what was in them; the chemist could now design entirely new materials and fashion the strategies to prepare them. The structural concepts of the aromatic benzene ring and the tetrahedral carbon atom provided a link between theoretical and applied chemistry; that is, they provided a knowledge of the structure of the desired products and the clues as to what transformations were needed to make the products. Some historians of science regard the late 1800s and early 1900s as the golden age of organic chemistry. Books popularizing science or chemistry for nonscientists often extolled coal tar with florid hyperbole, for example, "The Wealth of the Indes" and "The Magic Purse of Fortunates".

The demands for industrial chemicals eventually surpassed the production abilities of the coke industry. Beginning around 1925, petroleum-based chemicals became an increasingly important component of the chemical industry. At one time, about 95% of the phenol used in the United States was obtained from coal tar. Today, only about 5% of the phenol used is obtained from coal tar. However, this 5% represents about the same number of tons of coal-tar phenol as the 95% from the past. The phenol industry has greatly expanded, and other sources are now used to obtain this chemical. Benzene, toluene, and xylene—all coal-tar primaries—are also now obtained mostly from other sources.

Our modern organic chemical industry is largely a petrochemical, rather than coal chemical, industry. The future will represent a balance between the declining production of coal tar due to changes in the iron and steel industries and the prospect of shortages of liquid petroleum and natural gas. The use of coal chemicals may increase in the future, but most likely these coal chemicals will be produced from new processes rather than from coking byproducts.

Ammonia from Coal

Another useful chemical byproduct of coke manufacture is ammonia. The ammonia arises from reactions of organically bound nitrogen in the coal during carbonization. One ton of coal yields about 5–10 pounds of ammonia.

Recovery

Several processes are used to recover byproduct ammonia. The most common process in the United States is the *semidirect process,* in which ammonia is recovered as ammonium sulfate. The hot gases leaving the coke oven are cooled to condense the tar. The gas then passes through a dilute solution of sulfuric acid in which the ammonia reacts with the acid to produce ammonium sulfate. The solution can be evaporated to recover ammonium sulfate crystals. Alternatively, the solution can be treated with a base such as calcium hydroxide (lime) and heated to liberate the ammonia.

Free and Fixed Ammonia

Continued cooling and treatment of the gas stream from the coke ovens produce an aqueous gas liquor that also contains ammonia. The ammonia in the liquor exists in two forms: free ammonia and fixed ammonia. *Free ammonia* is the salt of a weak acid, such as ammonium sulfide or ammonium carbonate. *Fixed ammonia* is the salt of strong acid, such as ammonium chloride. Free ammonia can be released from the gas liquor by boiling it. Fixed ammonia can only be liberated by adding calcium hydroxide or some similar base before boiling. Fixed or free, the ammonia liberated from the gas liquor is collected by reaction with sulfuric acid.

Uses

Ammonium sulfate is used as a fertilizer. It is not a valuable fertilizer because it contains only about 25% of available plant nutrients. Sometimes ammonium sulfate will not sell at a price sufficient to cover the cost of the sulfuric acid used to make it. An alternative process uses phosphoric acid to capture the ammonia because the resulting ammonium phosphate salts contain nitrogen and phosphorus, which are both essential to plant life.

Ammonia is sold as a liquid compressed in tanks (anhydrous ammonia) or dissolved in water (ammonia water). Ammonia is a valuable intermediate in the production of sodium carbonate in the *Solvay process.*[1] Sodium carbonate is widely used in industrial inorganic chemistry, for example, in making glass and soap. The installation of the first byproduct coke ovens in the United States was motivated by the need for ammonia for the Solvay process rather than for the coke itself. In this case, the coke was the byproduct. Anhydrous ammonia can be directly applied to fields as a fertilizer. Ammonia was once very popular as the working fluid in refrigeration systems, but its use in this application has largely been supplanted by the commercial Freon refrigerants. The aqueous solution of ammonia is most familiar as household ammonia, which is actually a dilute solution of ammonia. Concentrated solutions of ammonia are used in industry for many of the same purposes as household ammonia, including the removal of grease and stains.

Ammonia, ammonium sulfate, and ammonium phosphate are the materials obtained directly from the coking plant. Most of the other ammonium compounds in commerce are made from one of these three and thus can be produced from coal. Ammonium chloride is a constituent of dry cells; ammonium bromide is used to make photographic paper and film; and ammonium nitrate doubles as both a fertilizer and an explosive.

Other Sources

Ammonia, like the coal-tar primaries, is now derived from other sources besides coal tar. Hydrocarbons, such as methane (in natural gas) react with steam to make carbon monoxide, carbon dioxide, and hydrogen. Subsequent absorption of the carbon monoxide and carbon dioxide leaves essentially pure hydrogen. The hydrogen can then be reacted with nitrogen to make ammonia. Today, about 96% of the ammonia produced is derived from natural gas and petroleum. In some cases, ammonia in a coking plant is more of a nuisance than a useful byproduct. In these cases, it is converted to nitrogen and water vapor by burning, and these products are released to the environment.

Future Production

Although ammonia produced from coal represents only a small fraction of total ammonia production today, renewed production of coal-derived ammonia is a likely possibility in the future. The need for ammonia, particularly for use as a fertilizer, is almost certain to increase. At the same time, however, a significant increase in coking capacity will not likely occur. The ammonia made from coal in the future will not be a coke byproduct but instead will come from coal gasification.

Synthetic Ammonia

Ammonia is synthesized by the *Haber process,* which was developed in Germany in 1913. In the Haber process, nitrogen and hydrogen react at 750–1100 degrees Fahrenheit and 200–700 atmospheres in the presence of an iron catalyst. The nitrogen is obtained relatively easily by distillation of liquefied air. The problem is to find a source of hydrogen. One route to hydrogen is the reaction of coal with steam and air or oxygen. The principal products of this gasification reaction are carbon monoxide and hydrogen. Separation of the gaseous products yields a stream of reasonably pure hydrogen for use in the ammonia synthesis. Some commercial gasification plants are already operating for the specific purpose of making hydrogen for ammonia synthesis. Most of these plants use the Koppers–Totzek gasifiers, which are described in the next chapter.

Acetylene from Coal

Uses

Calcium carbide is produced from coke by reaction with calcium oxide in an electric arc furnace. Calcium carbide is subsequently reacted with water to form acetylene. Acetylene is a gas widely used for the manufacture of monomers that are the starting materials in the manufacture of plastics. Acetylene is also used in the oxyacetylene torch for welding and cutting steel. When acetylene is reacted with hydrogen chloride, vinyl chloride is produced. The vinyl chloride is used to make a variety of plastics, particularly poly(vinyl chloride) (PVC), which is widely used today.

Transportation

Acetylene is produced from petroleum sources as well as from coke. Petroleum-derived acetylene has not competed with coal-derived acetylene for the market to the same degree that other petrochemicals have competed with coal chemicals. This situation stems from the difficulty of shipping large quantities of acetylene. Most industrial gases can be transported by liquefying them or compressing them at high pressures; the products are shipped in pressurized cylinders or tanks. Acetylene, however, explodes when compressed to more than twice the normal atmospheric pressure. The only safe way to handle acetylene at moderately high pressures is to dissolve it in acetone. The cylinder of acetylene on a welder's cart is actually a cylinder of acetylene dissolved in acetone. Industries that use large quantities of acetylene should be located near a source. The main disadvantage of using calcium carbide to make acetylene is the large amount of electrical energy needed to make calcium carbide. Making 1 ton of calcium carbide, which can eventually produce only 1500 pounds of PVC, requires the amount of electricity needed to keep a 100-watt light bulb burning continuously for 4 years.

Effect of Ethylene and Propylene

Acetylene is a versatile chemical that can be used to make other plastics such as poly(vinyl acetate), which is used in latex paints and plastic films, and acrylates, which are used in paints. Up to the 1940s, acetylene from calcium carbide was an important starting material for the production of many organic chemicals used in industry. After World War II, inexpensive ethylene and propylene became available in large quantities from the petroleum industry. Today, much of industrial organic chemistry is based on ethylene and propylene. Acetylene produced from calcium carbide is now virtually obsolete except for an operation in South Africa that produces PVC.

Future Production

In the future, supplies of petroleum will eventually dwindle, and acetylene produced from coal may once again be an important source of industrial chemicals. Most likely, however, calcium carbide will not be the intermediate between coal and acetylene. The large electric arc furnaces used for making calcium carbide would have to be scaled up to accommodate the tremendous tonnages of coal that would be required to meet the demand. Such a task would be difficult and costly. Consequently, new methods are being sought to produce acetylene from coal. An example is a plasma process being developed by AVCO. Finely pulverized coal is pyrolyzed in a *plasma* (a hot, ionized gas) at about 3600 degrees Fahrenheit. The coal is in contact with the hot gas for only a few thousandths of a second. The reaction products are rapidly cooled lower than 750 degrees Fahrenheit to slow down the decomposition of acetylene. This process is capable of converting 30% of the coal to acetylene.

Synthesis Gas

Another route to deriving chemicals from coal takes advantage of the chemical versatility of carbon monoxide. Reacting coal or coke with steam and air or oxygen produces a mixture of carbon monoxide and hydrogen called *synthesis gas.* Synthesis gas is a useful fuel; it has about one-third of the heating value of an equal volume of natural gas (methane). It can also be converted to methane and used as substitute natural gas.

Fischer–Tropsch Process

In the 1920s, two German scientists, Franz Fischer and Hans Tropsch, investigated the reactions of carbon monoxide and hydrogen in the presence of various catalysts at fairly high temperatures and pressures. The objective of this research was to produce light liquid hydrocarbons that could be used as gasoline. Fischer and Tropsch were successful in producing synthetic liquid fuels from carbon monoxide, and this development made an important contribution to the German fuel supplies during World War II. However, continued research showed that the *Fischer–Tropsch process* could yield a wide variety of products, depending on the catalyst and reaction conditions chosen.

Two versions of the Fischer–Tropsch process result in the production of straight-chain alcohols. The *synol process* occurs at 392 degrees Fahrenheit and 140 atmospheres and uses an ammonia catalyst. The *oxyl process* uses an iron catalyst at about the same temperature and pressure. The alcohols are a versatile family: The smaller alcohols are useful solvents, and the longer chain alcohols can be reacted with sulfuric acid to form alkyl sulfates, which are then neutralized with sodium hydroxide. The resulting sodium alkyl sulfates are synthetic detergents. Long-chain alcohols also serve as *emollients* (agents that soften or smooth the skin) in cosmetics.

Oxo Synthesis

The *oxo synthesis* uses much higher pressures (100–500 atmospheres) and somewhat lower temperatures (212–392 degrees Fahrenheit) than the Fischer–Tropsch process. This synthesis incorporates a cobalt carbonyl catalyst to react the carbon monoxide and hydrogen with olefins. The products of this reaction are aldehydes, which are converted to alcohols in a subsequent step. The alcohols are used in the same ways as those from the synthol and oxyl processes: (1) as solvents and (2) for making detergents, plasticizers, and cosmetics. The oxo process is an important industrial source of these materials; annual production is more than 2 million tons. The synthesis gas currently used does not come from coal but is made by reacting methane with steam, a process called *reforming*, or by reacting heavy petroleum oil with steam and air. The sources of carbon monoxide and hydrogen do not affect the reaction; therefore, as natural gas and petroleum supplies decline in the future, synthesis gas from coal can be used as the feed for the oxo process.

Other Reactions

Iron-based catalysts, temperatures of 392–572 degrees Fahrenheit, and pressures of 5–50 atmospheres produce a mixture of gasoline, diesel oil, and long-chain alkanes from carbon monoxide and hydrogen. During World War II, the long-chain alkanes that were too long to be used as liquid fuels were oxidized to fatty acids. These acids were a source of synthetic fats, for example, food fats such as margarine, which were in short supply in Germany during World War II.

Conclusions

This discussion of chemicals from coal began with an attempt to synthesize quinine. Quinine belongs to the class of compounds known as *alkaloids*; it has the most complicated molecular structure of any of the alkaloids, although Hofmann and Perkin were not aware of this fact. The total synthesis of quinine was achieved by another organic chemist, Robert B. Woodward of Harvard University. Woodward and his colleague William Doering reported the synthesis of quinine in 1944, 88 years after Perkin's attempt. In the meantime, superior synthetic antimalarial drugs, such as quinacrine hydrochloride, had become available. The quest for quinine led instead to dyes, explosives, perfumes, medicines, and flavorings; synthetic quinine is not made commercially even today.

Note

[1] The *Solvay process* is an important route to the production of sodium carbonate (also known as soda ash or washing soda) and sodium bicarbonate (baking soda). Deposits of these compounds are scarce and often far from markets. The basic reactions in the Solvay process are the reaction of ammonia and carbon dioxide with a saturated solution of sodium chloride to form sodium bicarbonate, which crystallizes out of the solution, and the subsequent heating of the bicarbonate to form sodium carbonate, carbon dioxide, and water. Ammonium chloride forms as the other product in the production of the bicarbonate; reacting ammonium chloride with calcium hydroxide (lime) regenerates the ammonia for recycling.

Chapter 10

Synthetic Fuels from Coal: Gasification

Straight from the ground, coal is a good source of energy. Untreated bituminous coal has a heating value of about 12,000–14,000 British thermal units per pound (Btu/lb), which compares quite well with heating values of 5600 Btu/lb for dry oak and 18,800 Btu/lb for aviation gasoline. Treatment of coal after removal from a coal seam and before use is an added expense that leads to higher energy costs. In this regard, why should we even consider converting solid coal to liquid or gaseous fuels?

First, the industrialized world has an incalculable investment in devices that obtain their energy from liquid or gaseous fuels, for example, the automobile and the gas stove. As supplies of petroleum and natural gas decline, the vehicles, power stations, and domestic appliances that consume liquid or gaseous fuels will not be easily replaced. Even if financial considerations are not a factor, the physical aspects of conversion or replacement will be overwhelming. Synthetic fuels from coal can provide a substitute for petroleum or gas to help bridge the transition from an industrial economy based on oil and gas to a new system using entirely different energy sources.

Second, coal contains impurities that are undesirable in systems that burn coal for energy. These impurities can be reduced or even eliminated while the coal is being converted to liquid or gas. Synthetic fuels burn more cleanly than coal; they form fewer sulfur and nitrogen oxides during combustion and have fewer ash-related problems.

Third, coal can be difficult and laborious to handle. At the mine it has to be loaded into rail cars or trucks. During transportation it can decrepitate, blow away, or undergo spontaneous heating. At the destination the coal has to be unloaded, moved around, and stockpiled for use. Somewhere along the line it may have to be moved through a beneficiation plant and be crushed or pulverized. In contrast, gases or liquids can be handled easily with a system of pipes, tanks, pumps, and valves. Synthetic fuels from coal offer the potential of much greater ease and convenience of handling. The major processes for making synthetic fuels from coal can be grouped into three categories:

- gasification,
- indirect liquefaction, and
- direct liquefaction.

Gasification converts coal into a gaseous fuel. *Indirect liquefaction* first gasifies the coal to a synthesis gas and then converts the synthesis gas to liquids. *Direct liquefaction* converts coal into liquid products without the intermediate step of gasification.

Direct and indirect liquefaction and gasification usually involve reacting the coal with some other material, such as steam or hydrogen. Carbonization or pyrolysis processes, which occasionally have been considered for making synthetic fuels, rely on heat to decompose the coal into gases or liquids.

Gas-Lighting

The least complicated gasification process is carbonization. In 1609, a Belgian chemist, Jan van Helmont, observed that gas was evolved when coal was heated. Eighty years later, John Clayton, a clergyman in Yorkshire, England, experimented with collecting gas from coal. Clayton filled inflatable bladders with the coal gas and demonstrated that the gas could be burned. A century after Clayton's experiments, William Murdock, a Scottish engineer, devised a system for lighting his home with gas. Murdock was an employee of James Watt, who is best known for improving the design of steam engines to make them practical power sources in factories and mines. Watt formed a partnership with Matthew Boulton to manufacture steam engines. In 1798, Murdock installed a gas-lighting system in the Boulton and Watt factory in Birmingham, England. Gas-lighting soon spread to other factories and mills in the Midlands.

Six years after Murdock's installation of lighting in the Boulton and Watt factory, Frederick Winsor obtained a British patent on the manufacture of coal gas. Winsor was a Moravian who had emigrated to England. In 1807, Winsor and his partners made the first public demonstration of the use of gaslights for street lighting. They illuminated Pall Mall in London, creating a sensation for the residents. Until this time, the street lights that were used provided inadequate illumination. Streets at night were hazardous because the darkness not only afforded a sanctuary for criminals, but it also made negotiating gutters and other obstacles difficult. The improvement afforded by gaslights was so dramatic that by 1816 most of London was lit by gas.

In the United States, David Melville, a resident of Newport, Rhode Island, followed Murdock's lead and illuminated his home with gas in 1806. The first gas company in the United States was formed in Baltimore in 1817 by an artist, Charles Peale. The use of gas for lighting was bitterly opposed by manufacturers of candles and by refiners of whale oil. However, their efforts were futile, and gas-lighting spread steadily throughout the country: Boston in 1821, New York in 1823, and Philadelphia in 1841.

Coal Gas

By the mid-1920s, about 20% of the gas supply in the United States was coming from coal. When energy became a public issue because of the oil crises of the 1970s, coal gasification was sometimes considered a new development that might reduce our dependence on oil and gas. Prior to World War II, at least 20,000 gasifiers were operated in the United States. Their demise was caused by the spread of natural-gas pipelines. Cheap natural gas from Texas reached the Midwest in the 1930s. The war effort delayed the expansion of pipelines; however, in the postwar years, natural gas took over the market for gas and thus eliminated the need for coal gasification, at least temporarily. The gas produced during carbonization has a variety of names: coal gas, town gas, city gas, or illuminating gas. It is produced by heating coal in the absence of air. The gas produced as a byproduct during coking can be distributed as a fuel if it has a market in the vicinity of the coke works. Selling the gas allows the coke manufacturer to get rid of it and provides some additional income. Alternatively, if the local market is mostly for gas rather than for coke, coal can be carbonized for the purpose of making gas. In this case, the gas is marketed aggressively, and the solid coke or char is viewed as the byproduct to be discarded or sold locally.

Carbonization Process

When gas is the desired product, coal is carbonized in vertical or horizontal retorts. Horizontal retorts are shaped like an inverted U; vertical retorts are generally rectangular and taper slightly toward the top. At relatively low temperatures, about 750 degrees Fahrenheit, carbonization yields considerable tar and not much gas. The gas, however, contains hydrocarbons, which produce a luminous flame when the gas is burned. At higher carbonization temperatures the gas yield is greater, but the gas contains fewer hydrocarbons. The general practice is to carbonize at the highest practical temperature to achieve the largest yield of gas possible. Hydrocarbon illuminants can be blended with the gas, or the gas can be burned on a mantle that becomes incandescent and supplies the illumination.

Impurities

Similar to what occurs in the coke-oven method, hot gas from the retorts is cooled to condense tars and an aqueous gas liquor that contains salts of ammonia and a variety of organic compounds. The chief impurities in the gas are carbon dioxide, ammonia, and sulfur compounds such as hydrogen sulfide. Most of the impurities are removed in scrubbers, where they are absorbed into a liquid. Sulfur compounds are particularly objectionable not only because of their odor but also because they form sulfur dioxide when the gas is burned. Sulfur dioxide can combine with moisture to form an acid mist that can deteriorate household furnishings. As a final cleaning step, the gas is passed over calcium hydroxide (slaked lime) or iron oxides to absorb the rest of the sulfur compounds.

Components

The main components of coal gas are hydrogen and methane. The composition of the gas varies depending on the specific coal being used and the carbonization conditions. Hydrogen constitutes about 40–50% of the gas, and methane constitutes about 30–40%. Other constituents, which might each be present in the range of 2–10%, are nitrogen, carbon monoxide, ethylene, and carbon dioxide. The amount of gas made per ton of coal also depends on the type of coal used and on the carbonization conditions. A typical yield is 10,000 cubic feet of gas per ton of coal carbonized. The heating value of coal gas is about 550–700 British thermal units per cubic foot (Btu/ft^3), whereas the heating value of natural gas is 1000 Btu/ft^3. For 1 ton of bituminous coal having a heating value of 13,500 Btu/lb, about 20% of the weight of the coal is converted to gas, and this amount of gas has about 23% of the heating value of the original coal.

Gaslight Era

Except for local use of byproduct gas from coke ovens, coal gas is no longer a commercial product. Even if declining supplies of natural gas lead to a revival of coal gasification, most likely carbonization will not be a competitive process because only a small percentage of the coal is converted to gas. However, the legacy of coal gas lives on in our language. The rapid spread of coal gas for lighting streets and homes in the 19th century led to the term *gaslight era,* which is often used to mean the late 1800s. The plant used to produce coal gas was often referred to as the *gashouse.* Carbonization produces a distinct odor, and in the days before the public showed much environmental concern, smelly and dirty byproduct tars and oils might be discarded near the neighborhood. The neighborhood and its inhabitants were fairly rough, and a group of roughnecks might be called the *gashouse gang.* This term was immortalized for sports fans by the St. Louis Cardinals baseball teams of the 1930s, who earned the name for their freewheeling and rough style of play.

Producer Gas

Although carbonization is a relatively simple process to perform, it has the great disadvantage of converting only a minor fraction of the coal to gas. Unless a market or use can be found for the byproduct coke or char, carbonization is an inefficient process for gasification. A process that converts virtually all of the carbon in the coal to gas would be preferable.

Combustion

A conversion process that consumes all of the carbon is combustion. Burning coal in an excess of air converts the carbon to carbon dioxide, which has no value as a fuel. If the amount of air is limited, the product is not carbon dioxide but is instead carbon monoxide. Carbon monoxide can be burned further, so it has potential use as a gaseous fuel.

Passing air slowly through a bed of hot coal converts most of the carbon to carbon monoxide. Any carbon dioxide that forms is, in principle, converted back to carbon monoxide by reactions with hot carbon. In practice, minor inefficiencies of operation result in some carbon dioxide escaping into the product gas. When bituminous coal is used, some inevitable carbonization of the hot coal adds a small percentage of hydrocarbons to the gas. The nitrogen in the air passes through the coal bed unreacted. The gas made by this method is called *producer gas.* Typically, it might contain 20–25% carbon monoxide, 55–60% nitrogen, 2–8% carbon dioxide, and 3–5% hydrocarbons.

Heating Value and Yield

Carbon monoxide and the hydrocarbons are good fuels, but they are diluted by the large amounts of nitrogen and carbon dioxide. Thus, the heating value of producer gas is only about 130–145 Btu/ft^3. The yield of producer gas is 150,000–170,000 cubic feet per ton of coal. For a bituminous coal having a heating value of 13,500 Btu/lb, about 5–6 pounds of producer gas is obtained per pound of coal, and about 70–90% of the heating value of the coal appears in the gas. Although the heating value of producer gas is only about one-fourth of the heating value of coal gas, the total energy in the producer gas is much greater because a much larger volume of gas is produced.

Uses

Although producer gas has a low heating value, it became a popular fuel in industrial applications because of its easy preparation. Producer gas has been used in various jobs in industry. It was frequently used for firing open-hearth furnaces in steel mills, and it was also used to fuel glass-making furnaces and pottery kilns.

In principle, any coal or coke can be used for making producer gas. Conversion to producer gas can be a potential use of low-grade coals that might be unattractive for other applications. Bituminous coal is used for making producer gas for open-hearth furnaces or other large furnaces. The hydrocarbon gases added from carbonization to the producer gas create a luminous flame. Luminous flames are advantageous in large furnaces in which the principal mechanism of heating is by radiant transfer of heat from the flame to the surroundings.

Producer gas is not an ideal fuel for domestic use because of its low heating value and high concentrations of poisonous carbon monoxide that make the gas potentially dangerous. The demise of open-hearth furnaces in the steel industry and developments such as furnaces fired by natural gas and electric heating have reduced the demand for producer gas in industry. Nevertheless, manufacturing producer gas is an efficient means of turning coal into gas.

Water Gas

Production

Steam reacts with hot carbon to produce a mixture of carbon monoxide and hydrogen. Because the gas is made from steam, the mixture is known as *water gas.* Water gas is made by igniting a bed of coal or coke and forcing air through it until the bed is white hot, at which time steam is passed through the mass of hot coal. The reaction that forms water gas is *endothermic*; that is, it absorbs heat as it proceeds. The coal bed cools progressively as the water gas is made, and it eventually becomes too cold (when the temperature drops lower than 1800 degrees Fahrenheit) for the reaction to proceed. Then, the steam must be turned off, and air must be blown through the bed to reheat the coal. The timing of the air–steam–air cycle varies, depending on the design of the plant, but the alternation between air and steam is accomplished in a matter of minutes.

Components

The reaction of steam with carbon should produce a gas containing 50% each of carbon monoxide and hydrogen. The leakage of air into the generator results in the formation of some carbon dioxide and the introduction of some nitrogen. A typical water gas contains 50% hydrogen, 40% carbon monoxide, and small amounts of carbon dioxide and nitrogen. The heating value of water gas is about 300 Btu/ft^3.

When a water-gas generator is being blown with air to reheat the bed, producer gas is made from the reaction of hot carbon with oxygen. One ton of coal yields about 35,000 cubic feet of water gas and 80,000 cubic feet of producer gas.

Addition of Hydrocarbons

The burning of both hydrogen and carbon monoxide produces a colorless flame. When gas-lighting was common, water gas had no use as an illuminant unless it was burned by a Welsbach burner. This type of burner contains a gauze mantle impregnated with thorium oxide and cerium oxide, and the gauze becomes incandescent when heated. An alternative is to add hydrocarbons to the gas mixture because they provide a luminous flame. One method for adding hydrocarbons is to spray the hot gas with oil. The heat thermally cracks the large hydrocarbon molecules in the oil to produce smaller hydrocarbons that remain in the gas. The product is referred to as *carbureted water gas,* which contains about 35% hydrogen, 30% carbon monoxide, 10–15% methane, 8–10% other hydrocarbons, and small amounts of nitrogen and carbon dioxide. An additional benefit of adding hydrocarbons is an increase in the heating value of the gas to about 530 Btu/ft^3.

Blending with Coal Gas

Water gas can be blended with coal gas. The mixture is sold as illuminating gas. The advantage of this blend is that the coke or char left over from making coal gas can be reacted with steam to make water gas. Together, the two processing steps consume all of the coal.

Uses

Water gas was once useful as a fuel or illuminant. Like producer gas, it was dangerous for domestic use because it contained carbon monoxide. The mixture of carbon monoxide and hydrogen is a useful starting material for synthesizing chemicals or liquid fuels in the Fischer–Tropsch process. Water gas is a good source of hydrogen. Treating water gas with steam oxidizes the carbon monoxide to carbon dioxide and increases the amount of hydrogen. (The conversion of carbon monoxide and steam to carbon dioxide and hydrogen is called the *water-gas shift reaction.*) Absorption of carbon dioxide leaves reasonably pure hydrogen. Alternatively, water gas can be used to reduce ferric oxide to iron metal and ferrous oxide. These two products can react with steam to make hydrogen. Hydrogen is in great demand in industry for processes such as the synthesis of ammonia and the hydrogenation of organic compounds. One example of the hydrogenation of organic compounds is the conversion of large molecules in petroleum to smaller molecules typical of gasoline, a process known as *hydrocracking.*

Gasification Process

The disadvantage of making water gas is the need to alternately blow air and steam through the coal bed. The air supplies heat from combustion to keep the coal hot enough to react with the steam. These two processes can be performed simultaneously. If a mixture of air and steam is reacted with coal, both gases can be injected continuously. The amounts of air and steam are balanced to provide just enough combustion to keep the steam–carbon reaction going. An air–steam mixture, or in some designs an oxygen–steam mixture, is the basis of a modern gasification process.

More than 100 schemes have been suggested for gasification, many of which exist only on paper. Some have been tested in a laboratory on a small scale involving only a few grams or pounds of coal. Some have been taken further to the pilot-plant stage for testing on a scale involving tons of coal. Few schemes have succeeded commercially.

Gasification processes are generally classified on the basis of the method used to bring the coal into contact with the gasifying medium, usually a steam–air or steam–oxygen mixture. A vertical bed of coal can be supported on a grate, and the gasifying medium is injected into the bottom. The product gases and the coal move in opposite directions; the gases rise through the bed while the coal descends as it is consumed at the bottom. Processes of this type are called *fixed-bed gasification*; however, this term is a misnomer because the coal bed moves through the gasifier as the gasification proceeds at the bottom and fresh coal is added to the top. Fluid-bed gasification is physically similar to fluid-bed combustion in that pulverized coal is fluidized by the gasification medium. Pulverized coal can also be suspended in the medium, and then the suspension can be blown through the gasifier. This operation is known as *entrained-flow gasification.*

Lurgi Gasifier

The most successful fixed-bed gasifier is the *Lurgi gasifier.* The Lurgi gasifier was developed in Germany during the 1930s. The gas produced is mostly a mixture of hydrogen and carbon monoxide, which is used as synthesis gas or is converted to methane (substitute natural gas).

Components

The basic components are a coal lock, a water-cooled pressurized reaction vessel, and an ash lock. The gasifier vessel contains a distributor, which evenly spreads coal across the surface of the bed, and a rotating grate. Some Lurgi gasifiers have been fitted with stirrers that break up coalescing lumps of caking coals.

Operating Conditions

Lurgi gasifiers typically operate at 30–35 atmospheres and use a steam–oxygen mixture. The elevated pressure and the use of oxygen instead of air represent trade-offs between economic and technical issues. For example, elevated pressure offers several advantages over atmospheric pressure:

- a higher rate of coal throughput for a gasifier of a given size;
- a greater amount of methane in the product gas, which is desirable when the final product of the plant is substitute natural gas; and
- a reduction or elimination of the need for costly compressors to raise the gas pressure to that of the gas-distribution pipelines.

On the other hand, pressurized gasifiers are more expensive to build than atmospheric pressure units of comparable size because they must be stronger. They also offer mechanical design problems for introducing the coal and withdrawing the ash. The use of oxygen requires investment in a separate plant for making the oxygen (usually by liquefying and then fractionally distilling air). However, when oxygen is used in a gasifier instead of air, the product gas is not diluted with the nitrogen of the air. The heating value of the gas from an oxygen-blown gasifier is therefore considerably higher than the heating value of an equal volume of gas from an air-blown unit. Processes that produce oxygen on a large scale were fundamental in the development of coal gasification methods that completely consume the coal and yield gas having a heating value comparable to that of coal gas.

Coal Treatment

A typical Lurgi gasifier is illustrated in Figure 10.1. During operation of a Lurgi gasifier, coal is charged to the coal lock. Because the gasifier is running continuously and at high pressure, the lock is sealed from the gasifier while a new charge of coal is added. Then the lock is sealed against the atmosphere, pressurized, and opened to the gasifier to allow coal to drop into the gasifier vessel. As the coal descends through the gasifier, it is

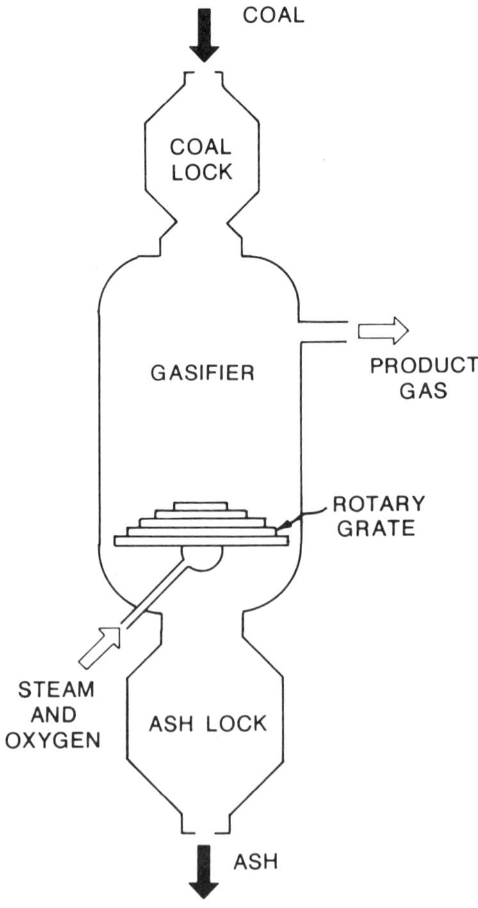

Figure 10.1. Although the Lurgi gasifier design is now about 50 years old, it is still the principal design used in commercial-scale synthetic fuel plants, including the Sasol plants in South Africa and the Great Plains plant in the United States. Coal descends through the shaft of the gasifier and reacts with a steam–oxygen mixture introduced through the grate. Gas and byproduct tars and oils are removed from the top of the gasifier; ash is removed from the bottom.

exposed to hot gases coming up through the coal bed. These hot gases first dry the coal. Then, as the coal descends further, it is carbonized and thus produces tar, gaseous hydrocarbons, and some carbon dioxide. The carbonized char descends further to react with the steam and oxygen.

Steam–Oxygen Reactions

The heart of the Lurgi gasifier is the grate. The steam–oxygen mixture is introduced through the grate. The oxygen reacts rapidly with the hot char to produce carbon dioxide. The oxygen reaction provides the heat needed for two reactions that occur higher in the bed: (1) steam reacting with char to produce carbon monoxide and hydrogen and (2) carbon dioxide reacting with the carbon in the char to produce carbon monoxide. Some hydrogen will react with the hot char to form methane. The gasification temperature is about 1800 degrees Fahrenheit.

Products

The raw gas leaving a Lurgi gasifier consists of the following:

- gasification products, namely, carbon monoxide, hydrogen, and methane;
- some carbon dioxide from carbonization reactions; and
- water from the moisture in the coal.

The raw gas is treated to remove the undesirable components such as tar and water. The cleaned gas product contains about 35% carbon monoxide, 50% hydrogen, and 15% methane and has a heating value of 475 Btu/ft^3.

The hot ash drops through the grate into the ash lock where it is cooled in water. Ash is periodically removed from the ash lock in a cycle of operation analogous to that of the coal lock: The ash lock is sealed from the gasifier, reduced to atmospheric pressure, opened for discharge, closed, pressurized, and then opened to the gasifier.

Commercial Units

Commercial Lurgi gasifiers are large units. Various size units are built, but a typical gasifier without the coal and ash locks is 12 feet in diameter and 20–25 feet high. Including the coal and ash locks, the overall height of the unit can be doubled easily. Gasifiers having various diameters are sold. A gasifier 12 feet in diameter can convert as much as 50 tons of coal to 1.6 million cubic feet of gas per hour.

Great Plains Plant

Planning and Construction

Lurgi gasifiers were used in the first commercial plant built in the United States for producing substitute natural gas. The plant was built by Great Plains Gasification Associates near Beulah, North Dakota. The project was begun by American Natural Resources about 15 years ago. It was originally envisioned to be a complex of four plants that would collectively produce 1 billion standard cubic feet per day of substitute natural gas. (*Standard* refers to a temperature of 32 degrees Fahrenheit and a pressure of 1 atmosphere.) This amount of gas represents about 40% of the gas consumed by the customers of American Natural Resources. After the corporate decision was made to proceed with the project, an incredible series of delays occurred because of legislative, regulatory, and bureaucratic processes. Construction did not start until 1980, by which time the project had been reduced in scope to constructing one-half of one of the four plants first planned, that is, a plant having a capacity of 125 million standard cubic feet per day of substitute natural gas. The plant serves as a commercial-scale gasification unit rather than a major supplier of substitute natural gas to the gas-distribution network.

Gasifiers

The heart of the plant is a train of 14 Lurgi gasifiers. The Lurgi gasifiers were chosen because of their reliability in commercial operations and probably because they were the best gasifiers available at the time the plant was designed. Actual operation of the plant uses only 12 of the gasifiers; therefore, at any given time, 2 of the 14 gasifiers can be out of service for maintenance.

Coal Use

When the plant is operating, about 22,000 tons of coal are delivered per day. The coal is separated by size. Pieces larger than one-fourth of an inch are retained for the gasifiers. The smaller pieces are sent next door to an electric power station owned by Basin Electric Power Cooperative. Basin Electric, in return for the coal, supplies electricity to the gasification plant. (Fixed-bed gasifiers cannot use fine particles of coal because the particles can settle through the larger pieces of coal and severely reduce the permeability of the coal bed to the gases flowing upward.) About 8000 tons of coal are provided to Basin Electric, and the remaining 14,000 tons are gasified. Each gasifier consumes about 500 tons of coal per hour.

Products

A block flow diagram of the processes in the plant is shown in Figure 10.2. The gasifiers operate at 29 atmospheres. In addition to the coal, the gasifiers consume about 3000 tons of oxygen and 14,000 tons of steam per day. The main product of the gasifiers is a raw gas in which the principal components are hydrogen and carbon monoxide. At this stage in the plant, the gas has a heating value of 310 Btu/ft^3. In addition to the gas, about 900 tons of ash are produced each day. The ash is returned to the mine for disposal in the pit.

The ultimate product desired is methane, which is formed in a methanation reaction of carbon monoxide with hydrogen. The raw gas does not have the appropriate ratio of hydrogen to carbon monoxide (3 to 1) for complete methanation in which three molecules of hydrogen react with one molecule of carbon monoxide to produce one molecule of methane and one molecule of water.

A portion of the raw gas reacts with steam to change the ratio of hydrogen to carbon monoxide. In this *shift reaction*, carbon monoxide reacts with steam to make carbon dioxide and hydrogen. The net effect of the shift reaction is to lower the carbon monoxide content and raise the hydrogen content. The shifted gas is then blended with the untreated raw gas, and finally the entire gas stream is cooled. Tar; oils; and an aqueous solution of ammonia, phenols, and several minor components condense as a result of the cooling process.

After removal of the condensate, the gas still contains carbon dioxide, sulfur compounds such as hydrogen sulfide, and some volatile aromatic compounds. These impurities are removed by contacting the gas with cold methanol, a procedure known as the *Rectisol process*. The purified gas then passes through methanation reactors containing a nickel-based catalyst. Finally, the gas is cooled, dried, and compressed to 100 atmospheres for feeding into the pipelines.

The substitute natural gas has a heating value of about 980 Btu/ft^3. If operated at full capacity, the plant would produce 137 million cubic feet of substitute natural gas per day. However, because the plant is assumed to be in full operation only 91% of the time, the average production is 125 million cubic feet per day. If the gas produced from the Great Plains plant were converted completely to electricity, it would produce about 1600 megawatts, which is the output of two large power stations.

Treatment of Waste Products

Environmental regulations require that the waste products be treated. The sulfur gases from the Rectisol process are converted to elemental sulfur. Hydrogen sulfide is oxidized to sulfur by sodium metavanadate. The reduced

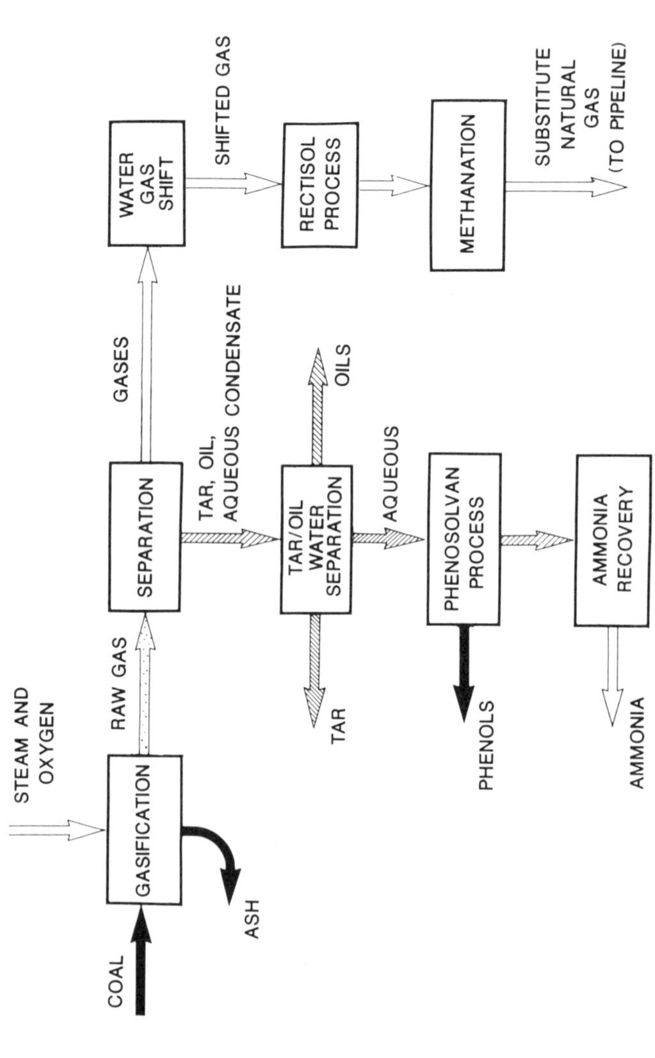

Figure 10.2. The Great Plains plant produces substitute natural gas by gasifying lignite, adjusting the gas composition by shifting, and then forming methane from the shifted gas. The byproducts from the plant have potential value as chemicals or fuels.

vanadium is converted back to the metavanadate form by reaction with air. The sulfur is recovered and sold as a byproduct. The commercial sodium metavanadate oxidation of hydrogen sulfide is called the *Stretford process*. The Stretford units produce 80–100 tons of sulfur per day.

Tar, Oil, and Water Separation

The tar, oil, and water mixture condensed from the raw gas is passed through separators that remove the oil and tar from the water on the basis of their specific gravities: The oil floats and the tar sinks in water. The water is then extracted with isopropyl ether to remove phenols and other organic compounds. The next step in water treatment is the removal of ammonia by the *Phosam process*, a procedure involving the absorption of ammonia in a solution of monobasic ammonium phosphate to form dibasic ammonium phosphate. The water is finally cycled through the plant's cooling towers. Water in which the remaining contaminants have been highly concentrated is discarded by deep-well injection. The daily recovery of ammonia from the Phosam process is about 90–120 tons. The ammonia is recovered by heating the Phosam solution to regenerate monobasic ammonium phosphate and ammonia. The ammonia is then sold as a byproduct, principally for use as a fertilizer. The phenols, tars, and oil are used as fuel to fire boilers and steam superheaters in the plant.

Total Production

From 14,000 tons of coal, 14,000 tons of steam, and 3000 tons of oxygen, the Great Plains plant produces 137 million cubic feet of substitute natural gas, 90 tons of sulfur, 105 tons of ammonia, and 900 tons of ash as waste.

Setbacks

Construction of the Great Plains plant began in October 1983, and full plant operation was achieved in July 1985. On the basis of its operation since July 1985, the plant has achieved reasonable technical success. Unfortunately, because of the current economic situation, the future of the plant is uncertain.

In 1975, the plant was predicted to be in full operation in October 1981 and sell gas at a cost of $3.00 per thousand cubic feet (in 1985 dollars). After 4 years of regulatory and legal snarls, full gas production was predicted for December 1983. The cost of the plant was expected to be $900 million. By 1981, the plant was expected to operate fully in December 1984, and the estimated price of the gas had risen to $6.75 per thousand cubic feet. Although the predicted gas cost had more than doubled, the price still seemed attractive in the context of 1981 energy costs. For example, an equivalent amount of energy in the form of No. 2 fuel oil (an oil widely used for domestic heating) was $7.25. The plant has now been built at a cost of $2.1 billion, and the price of the gas is in the range of $9.00–$10.00 per thousand cubic feet. However, because of the recent sharp decline in energy costs, natural gas is being pumped into pipelines at $2.50–$3.00 per thousand cubic feet. The Great Plains plant is producing substitute natural gas at triple or quadruple the cost of natural gas. No plant of any kind can operate for long under these conditions without some form of price support or subsidy.

Lessons for the Future

Two lessons emerge from the Great Plains experience. First, in a time of rapidly changing energy prices, tremendous risks are involved in deciding to build a plant, especially when one considers the time required for the regulatory permitting process and the actual construction. The decision to build even a small synthetic fuels plant involves a commitment of $1 billion. This amount may seem a wise investment when the projected cost of the synthetic fuel is lower than or even roughly equivalent to the cost of natural fuels. However, because of fluctuating energy costs, the same investment will appear very foolish if the fuel produced by the finished plant costs substantially more than natural fuel. Few corporate managements have the financial base or the unerring accuracy of predictions to bet $1 billion on the cost of gas or oil 15 years from now. For this reason and because of the general decline in energy costs, many corporations have been reducing or eliminating their involvement in research and development of synthetic fuel production.

The second lesson learned from the Great Plains situation involves a consideration of the consequences of not building synthetic fuel plants, although the technology is available; the Lurgi gasifier is a prime example of this technology. Some people argue that synthetic fuel plants should not be built now because current supplies of natural and imported fuels are adequate. Instead, the plants can be built when supplies of natural fuels decline substantially or when supplies of imported fuel are cut off. On the basis of the Great Plains experience, if a severe disruption occurred in the Middle East today, the lost imported fuel could be replaced with synthetic fuel in about 15 years. However, in the case of a serious national emergency, the bureaucratic and legal delays endured by Great Plains would probably be short-circuited or removed. Nevertheless, the plant engineering and construction would still take years, and these years could be ones of serious political and economic strain. This concern raises the counterargument that synthetic fuel plants should be built now, even if price supports are necessary, so that the plant capacity and the operating experience will be available when they are needed. The resolution of this argument cannot be accomplished by coal researchers; instead, the decision will evolve from the complex interactions of politics, economics, and social concerns.

Fluidized-Bed Gasifiers

Winkler Gasifier

One type of fluidized-bed gasifier that has reached commercial operation is the Winkler gasifier. The Winkler gasifier was a developed in Germany in the late 1920s. The impetus for designing and developing a fluidized-bed gasifier was to obtain a system for making synthesis gas from coals having a small particle size (less than 10 mm) or from coals that are too *friable* (easily crumbled or crushed into powder) for use in the existing fixed-bed gasifiers.

A typical Winkler gasifier (Figure 10.3) is about 75 feet high and 15–18 feet in diameter. Coal is fed to the gasifier by a screw feeder. If producer gas is the desired product, an air–steam mixture is blown through the grate at atmospheric pressure to fluidize the bed and carry out primary gasification.

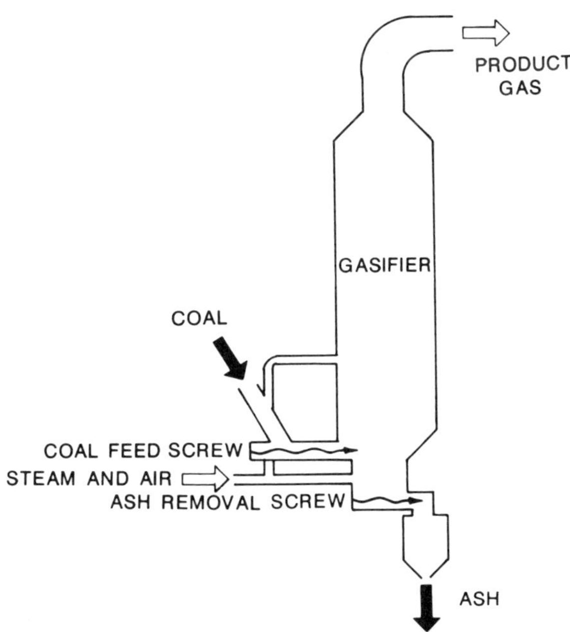

Figure 10.3. The Winkler gasifier uses a fluidized bed to effect reaction among the coal, steam, and air. The coal is fed to the gasifier, and ash is removed by screw feeders. Winkler gasifiers have been used commercially.

If synthesis gas is the desired product, an oxygen–steam mixture is used. A secondary blast is introduced above the bed to burn or gasify char or tar escaping with the product gas. The synthesis gas will contain about 45–47% carbon monoxide, 35–37% hydrogen, 16% carbon dioxide, and small amounts of methane. Ash is removed through the bottom of the gasifier by a rotary scraper. A Winkler gasifier consumes about 1000 tons of coal per day, which produces about 40 million standard cubic feet of gas.

The operating temperature of the Winkler is 1800 degrees Fahrenheit. At atmospheric pressure, the gasification of high-rank coals is slow at this temperature; therefore, low-rank coals are a preferred feed. The ash should not sinter or melt in the bed because the bed can agglomerate and lose the ability to fluidize. Therefore, lignites or subbituminous coals that have high ash fusion temperatures are ideal coals for the Winkler gasifier.

Approximately 40 Winkler units are used commercially, mainly in the Near East and eastern Europe. Rheinische Braunkohlenwerke in Cologne is developing a pressurized Winkler that will operate at 10 atmospheres and temperatures higher than 2000 degrees Fahrenheit. This pressurized unit will have a higher efficiency than the current atmospheric pressure gasifiers.

Westinghouse Gasifier

A fluidized-bed gasifier developed by Westinghouse has been tested successfully since the late 1970s in a large-scale (1200 pounds of coal per hour) pilot plant; however, it has not yet achieved commercial status. The objective of the Westinghouse program originally was to produce a gas of low heating value, usually less than 300 Btu/ft^3, for electric power generation. The Westinghouse gasifier consists of two fluidized beds: a devolatizer–desulfurizer and a combustor–gasifier. The Westinghouse gasifier operates at pressures up to 20 atmospheres.

Dried coal is fed to the devolatizer–desulfurizer and is fluidized with the raw gas from the combustor–gasifier. The bed in the devolatizer–desulfurizer is recirculated rapidly to reduce agglomeration of caking coals and to reduce the temperature of the fluidizing gas so that methane formation is increased. The desulfurization is done by adding dolomite. After reaction with the sulfur compounds, the dolomite is regenerated by reaction with steam for eventual recycling to the devolatizer–desulfurizer. The gas made in the devolatizer–desulfurizer is the product of the system; it contains nitrogen, hydrogen, carbon monoxide, and carbon dioxide and has a heating value of 150 Btu/ft^3.

The solid product of the devolatizer–desulfurizer is a char that is fed to the gasifier–combustor unit. The gasifier–combustor is blown with an air–steam mixture. The operating temperatures are 1800–2150 degrees Fahrenheit; some agglomeration of the ash can occur in this temperature range. The agglomerated ash is removed from the bottom of the gasifier–combustor. The raw gas is sent to the devolatizer–desulfurizer. The

Westinghouse gasifier can use caking coals because the rapid recirculation in the devolatizer–desulfurizer prevents coal particles from caking. Using an oxygen–steam mixture in the gasifier–combustor results in a product gas that has a heating value of about 270 Btu/ft^3.

Entrained-Flow Gasifiers

Koppers–Totzek Gasifier

The Koppers–Totzek gasifier, developed in the 1930s, has been the most successful entrained-flow gasifier. More than 50 Koppers–Totzek gasifiers are being used commercially around the world. Finely pulverized coal is blown, with steam and oxygen, into diametrically opposed burners. A Koppers–Totzek gasifier containing two burners (frequently referred to as a two-headed gasifier) is about 25 feet long, 12 feet in diameter at the widest part, and 8 feet in diameter at the tapered ends. Koppers–Totzek gasifiers operate at atmospheric pressure. Four-headed gasifiers are also used; these have about twice the capacity of the two-headed units. Two-headed gasifiers used in South Africa consume about 10 tons of coal per hour, which produces about 650,000 cubic feet of synthesis gas.

Because the coal is blown rapidly into the gasifier, the residence time is relatively short, that is, about 1 second. (A coal particle might spend tens of minutes passing through a fixed-bed gasifier.) The short residence time means that the individual coal particles must be very small to react completely. The feed coal is pulverized to sizes less than 0.004 inches (less than 0.1 mm). Intense temperatures are generated; they range from 2750 to 2900 degrees Fahrenheit in the center of the gasifier to 3600 degrees Fahrenheit at the hottest point of the flame. These temperatures are higher than ash fusion temperatures; therefore, the ash melts to a slag, which runs down the gasifier and is quenched in water. Some of the slag may escape as a mist in the product gas; it condenses to a dust that must be removed from the gas. The product gas consists of about 55–60% carbon monoxide, 30% hydrogen, and 10% carbon dioxide. This carbon dioxide concentration is low compared with the amount produced by many other gasifiers.

A great advantage of the Koppers–Totzek gasifier (Figure 10.4) is its ability to use any kind of coal as long as the coal can be finely ground. The high temperatures of the gasifier ensure rapid rates of gasification of all ranks of coal and ensure the melting of any type of ash. About 99% of the carbon in the coal is converted to gas. Another advantage is that no tars or waste waters are produced as byproducts; therefore, the need for waste-treatment operations, such as those used in Lurgi plants, are eliminated.

The Koppers–Totzek gasifiers also have several disadvantages. The product gas may leave the gasifier at 2500 degrees Fahrenheit. In current

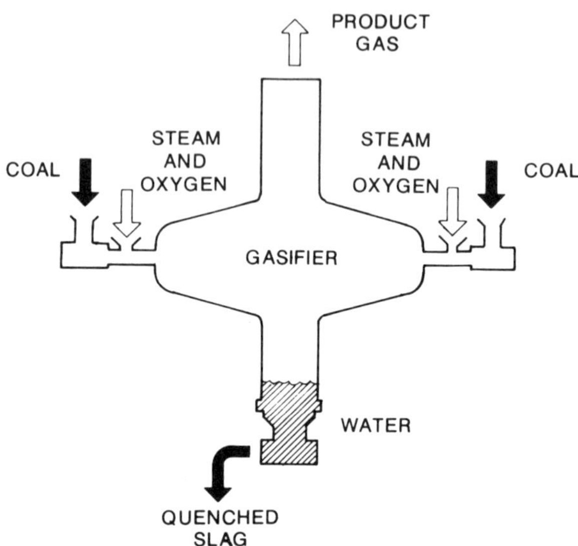

Figure 10.4. The Koppers–Totzek gasifier operates at atmospheric pressure and at temperatures high enough to melt the coal ash to a slag. These gasifiers are used commercially to make hydrogen for subsequent synthesis of ammonia fertilizers. This illustration shows a two-headed gasifier; four-headed versions are also built.

Koppers–Totzek plants, the *sensible heat* (i.e., the total heat content of the hot product gas relative to some selected reference temperature) is wasted when the gas is cooled. Because the gas is at atmospheric pressure, it must be compressed for subsequent use; gas compression adds the costs of compressors and their operation to the plant costs. The oxygen consumption is about 2.5 times the amount used for gasifying a comparable amount of coal in fixed-bed or fluidized-bed units.

Shell–Koppers Gasifier

Current research by Shell is directed toward the development of a pressurized version of the Koppers–Totzek gasifier. The Shell–Koppers unit operates at about 30 atmospheres and includes design features to generate steam by recovering the heat from the product gas. Pressure operation reduces the need for subsequent gas compression. The Shell–Koppers gasifier is now being tested on a pilot-plant scale.

Texaco Gasifier

Another type of entrained-flow gasifier nearing commercial status is the Texaco gasifier (Figure 10.5). This unit is an adaptation of a process

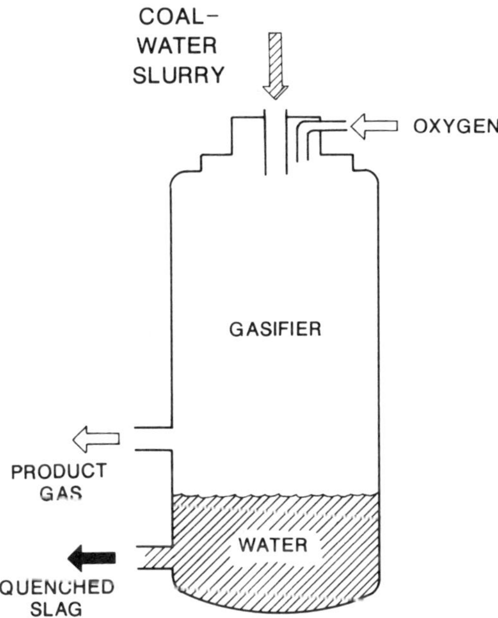

Figure 10.5. In the Texaco gasifier, a slurry of coal in water is entrained in oxygen and fired downward. Reaction temperatures are high enough to melt the coal ash. The Texaco gasifier is likely to be the gasifier used in future combined-cycle plants.

developed previously by Texaco for gasification of heavy fuel oils. It operates at pressures up to 80 atmospheres. The temperatures in the reaction zone range from 2000 to 2500 degrees Fahrenheit. They are controlled to be higher than the ash fusion temperature and to maintain a reasonably low viscosity of the resulting slag so that the slag will drain easily from the gasifier. The product gas contains about 45–50% carbon monoxide, 35% hydrogen, 15% carbon dioxide, and small amounts of methane. The Texaco gasifier, like the Koppers–Totzek gasifier, does not produce byproduct tars.

The Texaco gasifier operates by feeding a slurry of coal in water to the reactor. The slurry is preheated to about 1000 degrees Fahrenheit and is pressurized to 15 atmospheres. Preheated oxygen is blown into the gasifier to react with the coal–water slurry.

A persistent problem in the design of coal gasifiers is how to feed coal into a vessel that is at elevated pressure. One approach has been to use lock hoppers, which are used on the Lurgi gasifier. Lock hoppers can have mechanical problems, particularly around the valve; the valve must consistently and repeatedly produce a tight seal against the gasifier vessel as moderately abrasive and gritty coal passes through it. The use of slurry feeding is an alternative approach; its advantage is that raising the pressure of a liquid and pumping it into a pressurized vessel is mechanically simpler than a lock-hopper system.

Lignites can contain more than 35% moisture; as a result, slurrying a lignite with water results in a mixture containing too much water to be evaporated by the heat generated in a Texaco gasifier. However, scientists have possibly found a way to dry lignites so that they do not reabsorb moisture, even in a water slurry. A slurry of 65% dry coal in water contains 35% water and therefore has the same heating value as an untreated coal that contains 35% moisture. These findings indicate that slurries of dry lignite may be suitable feeds for Texaco gasifiers.

Considerable pilot-plant testing has been done with Texaco gasifiers in Montebello, California, and Oberhausen, West Germany. Commercial-scale units have been used by Olin Matheson to produce gas for ammonia synthesis and by Tennessee Eastman to produce synthesis gas for chemical manufacture. In the future, the Texaco gasifier may be used in combined-cycle plants[1] for electricity generation. A gasifier that uses 1000 tons of coal per day is being tested in this application at Southern California Edison's Cool Water station.

Conclusions

In the mid-1980s, ample supplies of natural gas appear to be available for the near future. Currently, coal gasification to produce substitute natural gas is not needed, as shown by the economic problems of the Great Plains plant in North Dakota. However, although natural gas is plentiful, other opportunities and needs now exist for coal gasification.

Small-scale gasifiers can be used to generate gas that has a relatively low heating value but is useful for process heat or steam generation in factories or industrial parks. Gasifiers can produce synthesis gas for the production of ammonia or organic chemicals. Electric utilities are showing steadily increasing interest in combined-cycle generating plants, which offer a promising future for gasifiers. By the end of the 20th century, the most important product of a coal gasifier may be electricity rather than gas.

Note

[1] Combined-cycle plants represent a potentially major contribution to future electricity needs. They are discussed more fully in Chapter 7.

Chapter 11

Synthetic Fuels from Coal: Liquefaction

Processes for making liquid fuels from coal can be grouped into two categories: indirect liquefaction and direct liquefaction. In *indirect liquefaction,* coal is first gasified to a mixture of carbon monoxide and hydrogen, that is, synthesis gas; then, the synthesis gas is converted to liquid hydrocarbons. This is an indirect process because the liquid product is formed from the synthesis gas rather than from the coal. In *direct liquefaction,* however, coal is reacted with hydrogen or hydrogen-rich organic compounds to produce a liquid fuel directly from the coal without the intermediate gasification step. Although direct liquefaction seems more straightforward and therefore more technically and economically desirable than indirect liquefaction, the world's largest complex for producing synthetic liquid fuel is based on indirect liquefaction.

Fischer–Tropsch Process

The most extensively studied and used indirect liquefaction method is the Fischer–Tropsch process, which was developed in Germany in the 1920s by Franz Fischer and his colleagues, most notably Hans Tropsch. The basis of the Fischer–Tropsch process is the hydrogenation of carbon monoxide in the presence of a catalyst. The hydrogen and carbon monoxide mixture is obtained from gasification. The products of the Fischer–Tropsch process are determined by the temperature, pressure, and catalyst selected for the reaction. The products, depending on the specific conditions selected, can include straight-chain alkanes and alkenes, alcohols, aldehydes, ketones, and fatty acids.

Operating Conditions

Fischer's original goal was to produce light hydrocarbons that could replace petroleum-derived gasoline. Research in the early 1920s used iron-based catalysts and reaction conditions of 750–850 degrees Fahrenheit and 100–150 atmospheres. Operating at these conditions yielded a mixture of alcohols, aldehydes, acids, and hydrocarbons; Fischer called this mixture *synthol.* Tests at much milder conditions (450–570 degrees Fahrenheit and 1 atmosphere) increased the proportion of hydrocarbons in the product mixture but lowered the rate of formation of products. This new problem stimulated research on the formulation of catalysts that would produce acceptable rates of hydrocarbon formation.

Octane Ratings

The straight-chain hydrocarbons, which are the usual hydrocarbon structures produced in a Fischer–Tropsch process, have relatively low octane ratings and therefore are less desirable as motor fuels than branched-chain hydrocarbons. Several approaches have addressed this problem, including varying the reaction conditions to allow the straight-chain compounds to rearrange to branched chains, treating the Fischer–Tropsch products in petroleum refinery operations designed to convert straight chains to branched chains, and modifying the reaction to produce branched-chain hydrocarbons directly.

Use in Germany

The Fischer–Tropsch process was a vital component of the German economy during World War II. Nine Fischer–Tropsch plants produced about 200 million gallons of fuel per year at their peak of production in 1943. Some oxygenated compounds were also produced, and these became important sources of fats, detergents, and soaps for domestic consumption. The synthesis was operated in the presence of cobalt-based catalysts partly because cobalt was not as important strategically as iron and nickel were. Some of the synthesis gas was obtained from the gasification of coke. Albert Speer, in his memoir *Inside the Third Reich,* describes the severe bombing of these synthetic fuel plants on May 12, 1944, as the day the outcome of the "technological war" was settled[1]. Japan also used synthetic fuel produced by the Fischer–Tropsch process, although not to the same extent as Germany. Four plants in Japan provided about 80 million gallons of fuel in 1942, the best year of production.

After World War II, synthetic liquid fuels could not compete economically with natural petroleum. In Europe, gasoline and chemicals could be obtained more cheaply from Middle Eastern oil. In the United States, the postwar development of oil fields in the Southwest provided a cheaper source of fuel.

Oxo Synthesis

Research and development of the Fischer–Tropsch process has continued and has been spurred mainly by the chemical industry. A variant of the Fischer–Tropsch process, called the *oxo synthesis,* reacts synthesis gas with olefins to form aldehydes, which are converted to solvents, plasticizers, or detergents in subsequent processing steps. The oxo synthesis is used extensively in the United States and Europe, although all of the synthesis gas comes from gasifying oil or from the reaction of steam with methane.

Medium-Pressure and Iso Synthesis

Two versions of the Fischer–Tropsch process are potentially useful for producing liquid fuels. The *medium-pressure synthesis* uses an iron-based catalyst and operates at 425–650 degrees Fahrenheit and 5–50 atmospheres. The products are a range of hydrocarbons ranging from gasoline to heavy paraffins. The relative amounts of the products can be changed by altering the ratio of hydrogen to carbon monoxide in the synthesis gas. Increasing the relative proportion of hydrogen increases the amount of gasoline in the product. The other version is the *iso synthesis*; this method uses thorium oxide as the principal catalyst and operates at 750–950 degrees Fahrenheit and 100–1000 atmospheres. The iso synthesis produces mainly low-molecular-weight branched-chain hydrocarbons. At the higher end of the temperature range, some aromatic hydrocarbons are produced as well. The iso synthesis products are useful for increasing the octane number of fuels.

Sasol Plants

Although synthetic liquid fuels produced by the Fischer–Tropsch process are not now economically competitive with petroleum-derived fuels, other factors besides economics can strongly influence the energy policy of a nation. For example, in Germany during World War II, the military need for fuel far outweighed cost considerations. The current situation in South Africa represents another example. South Africa is well-endowed with coal deposits but lacks significant petroleum resources. South Africa's notorious domestic racial policy of apartheid makes the nation susceptible to oil embargoes and protracted military struggle, either internally or against the black nations on its border. Thus, national security and strategic considerations have made the conversion of the readily available low-cost coal to liquid fuels an attractive option.

Planning and Construction

In 1951, South African officials decided to proceed with the first synthetic liquids plant. Fischer–Tropsch technology was selected because of its successful use in Germany before and during World War II and because direct liquefaction had not been used on the scale envisioned by the South Africans. The plant was ordered by the South African Coal Oil and Gas Corporation, which is generally known as Sasol. Construction of the Sasol plant began in 1953 at Sasolburg, outside Johannesburg, and the first oil was produced in 1955. This achievement is impressive and is particularly noteworthy when compared with the delays endured by the Great Plains gasification plant in the United States prior to its operation.

Gasifiers

The heart of the Sasol plant is a battery of Lurgi gasifiers. Originally, the plant consisted of nine gasifiers that produced 10 million cubic feet of raw gas per day. By 1975, the plant consisted of 13 gasifiers that produced about 23 million cubic feet of gas from 10,000 tons of coal each day.

Fischer–Tropsch Reactors

In the original Sasol plant the gas was split between two types of Fischer–Tropsch reactors. The majority was fed to the *Arge process,* which reacted the gas over an alkaline iron catalyst at 20 atmospheres to produce a mixture of gasoline, diesel fuel, heating oil, and waxes. The remainder was reacted in the synthol process in the presence of a reduced form of magnetite as the catalyst. The synthol reactor yielded gasoline, heating oil, and low-molecular-weight hydrocarbons; the low-molecular-weight hydrocarbons were sold as liquefied petroleum gas, which is known as LPG. A variety of alcohols and other chemicals were obtained as byproducts, all of which were sold.

Products

The ideal gasifier for preparing feed to a Fischer–Tropsch reactor is one that produces a gas composed mainly of carbon monoxide and hydrogen and containing minimal methane and few other byproducts from the gasification step. Lurgi gasifiers do not meet this ideal description. Nevertheless, Lurgi gasifiers were selected for the Sasol plant because they are a known commodity; that is, they are known to operate successfully on a large scale and therefore provide less technological risk, even though byproducts are produced. The methane is recovered and sold in the Sasolburg vicinity; it is blended with some unreacted gas from the Fischer–Tropsch units. The gas has a heating value of 650 British thermal units per cubic foot (Btu/ft^3). This byproduct gas is estimated to save South Africa the cost of importing 500,000 tons of oil per year. Other gasification byproducts are treated and recovered in operations similar to those described for the Great Plains plant. (The Sasol experience was a useful guide for some of the design elements of the Great Plains plant.) Treatment of the waste water and tars allows recovery of phenols, creosote oil, and a heavy pitch. Sulfur and ammonia are also recovered and sold.

The Sasol plant begins with coal and produces LPG, gasoline, diesel oil, heating oil, various chemicals, gas for residential heating, ammonia, sulfur, phenols, creosote, and pitch. Most of these products are important to the South African economy. Figure 11.1 shows a block flow diagram of Sasol operations. The Sasol products are reminiscent of the early days of the meat-packing industry in the United States when the industry boasted that it used every part of the pig but the squeal.

Sasol–2 and Sasol–3

Operation of the original plant continued steadily into the 1970s. The oil price shock of 1973 prompted a decision to proceed with a second plant called Sasol–2. This plant was constructed in Secunda, a new town that was built to support the plant and is 50 miles from Sasolburg. Sasol–2 is about triple the size of what is now called Sasol–1. Other than the scale of operation, Sasol–2 differs from Sasol–1 in two respects: (1) All of the synthesis gas is fed to synthol reactors because the principal desired product is gasoline. (2) The methane is reacted with steam to produce more carbon monoxide and hydrogen because methane has no market in the vicinity of Sasol–2. Sasol–2 consumes about 38,000 tons of coal per day and produces 49 gallons of gasoline from each ton. The Sasol–2 plant became operable in 1981.

While Sasol–2 was under construction, the decision was made to proceed with Sasol–3, a duplicate of the second plant. When the three plants are in operation, they will collectively supply synthetic fuels equivalent to about 70% of South Africa's need for crude oil.

Technical Success

Whether or not the Sasol operation would be judged successful solely on economic criteria is not clear. In the mid-1970s, when crude oil was selling at $11–$12 per barrel, Sasol–2 would have been considered barely profitable

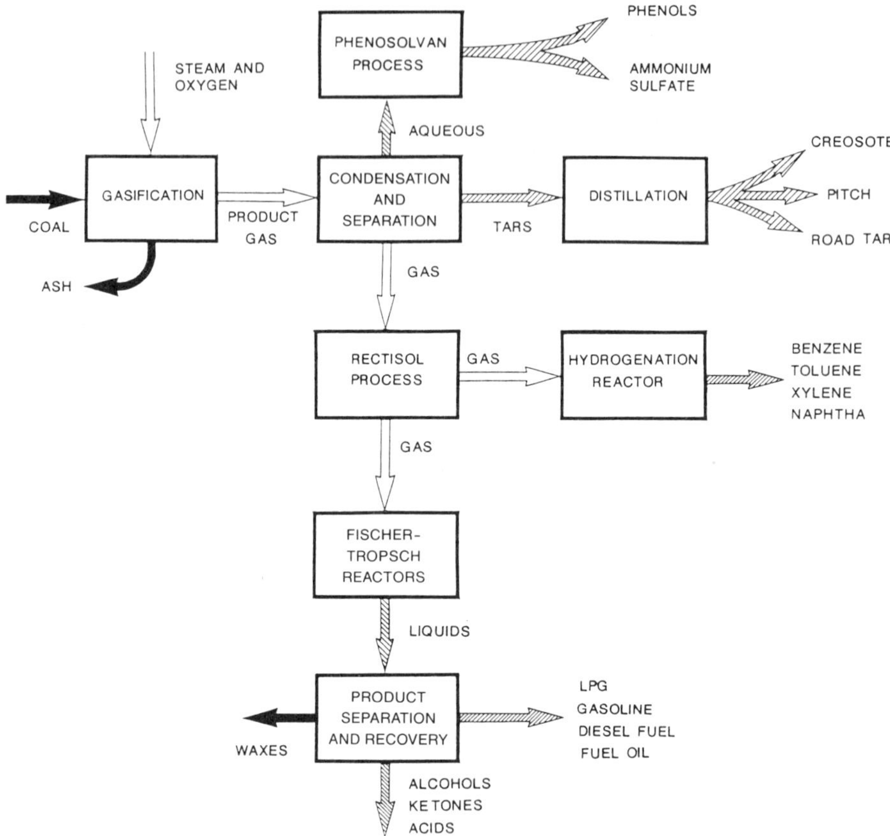

Figure 11.1. The production of synthetic liquid fuels at the Sasol plants involves gasification followed by Fischer–Tropsch synthesis. The other process streams in the plant are used as sources of a variety of chemicals.

on the basis of a mine-mouth cost of coal in the range of $4–$5 per ton. However, in a completely free economy, that same coal could have been exported at triple the price. Nevertheless, the technical success of Sasol is unquestioned, and South Africa is in a situation in which economics cannot be the only basis for decisions about synthetic fuel production.

Bergius Process

Direct liquefaction processes take an entirely different approach. Petroleum contains two hydrogen atoms for each carbon atom; coal contains about three hydrogen atoms for every four carbon atoms. Converting coal into a petroleumlike liquid therefore requires adding hydrogen to the coal to raise the hydrogen-to-carbon ratio from 0.75 to 2.0. Direct liquefaction processes are often referred to as *coal hydrogenation.*

Background

The first significant work on direct liquefaction was done about the time of World War I by the German scientist Friedrich Bergius. Bergius is noteworthy for being the only coal scientist ever to win a Nobel Prize (for chemistry in 1931). Bergius did not begin his research program by trying a variety of reaction schemes to hydrogenate coal. Instead, using coal samples obtained from all over the world, he spent at least 2 years in fundamental research examining the physical properties and chemical constitution of coal. To gain a further understanding of coal, Bergius also conducted experiments on artificial coalification, that is, reactions by which wood (sawdust) is converted to coal-like materials. Then, having a good working knowledge of the composition, properties, and formation of coal, Bergius proceeded to examine hydrogenation chemistry.

Reaction Conditions

Several factors influence the choice of reaction conditions for direct liquefaction:

1. Coal shows little reactivity toward hydrogen at temperatures lower than 650 degrees Fahrenheit, a temperature at which thermal decomposition of the coal structure begins to occur.
2. High pressures of hydrogen help drive the reaction toward higher conversions.
3. Fine particles of coal are more reactive than larger particles.
4. Coal slurries are fed more easily to a reactor than are solid materials.
5. Catalysts increase the rate of reaction.

On the basis of these factors, various reaction conditions have been selected for direct liquefaction by the Bergius process.

Products

In general, the Bergius process converts 1 ton of coal to 40–45 gallons of gasoline, 50 gallons of diesel fuel, and 35 gallons of fuel oil. The gasoline fraction contains 75–80% paraffins and olefins and 20–25% aromatic compounds. The relatively large percentage of aromatic compounds adds desirable antiknock qualities; the octane number[2] of the gasoline is 75–80.

(For comparison, unleaded regular gasolines have octane ratings in the high 80s.)

Middle and Heavier Oils

The process produces gasoline and other fuels in two stages: First, the coal is hydrogenated in the liquid phase to produce a *middle oil* that boils between 350 and 620 degrees Fahrenheit (roughly comparable to No. 2 domestic fuel oil, of which 90% must distill between 540 and 640 degrees Fahrenheit). Then, in the second stage, the middle oil is hydrogenated further in the vapor phase to gasoline and diesel fuel. In a large-scale operation, low-rank coals are hydrogenated at 900 degrees Fahrenheit and 240–300 atmospheres in a series of four reactors, each about 60 feet tall and 3 feet in diameter. A commonly used catalyst is ferric oxide in the form of red mud, a byproduct of the aluminum smelting industry. Unlike most catalysts, red mud is cheap and abundant enough to be used in large quantities and thrown away after use. Ferrous sulfate is also used as a catalyst. About 50% of the coal is converted to middle oil, and another 40% is converted to a heavier oil. The heavier oil is recycled through the reactor by being used as the vehicle for slurrying the coal. For bituminous coal liquefaction, hydrogen pressures of 340–700 atmospheres are used, and stannous oxalate or ferric oxide serves as the catalyst. A 60% conversion to middle oil can be achieved. The vapor-phase hydrogenation of the middle oil is carried out at 300 atmospheres and uses a tungsten disulfide catalyst.

Commercial Development

The first commercial plant was built in Leuna, Germany, in 1927 by IG Farben. Additional plants were erected during the 1930s; therefore, synthetic fuels from direct liquefaction also contributed to the war effort. Although estimates of the total synthetic fuel production in Germany and its allies during World War II vary, the amount could have been approximately 1 billion gallons per year. A plant was also built in Great Britain; however, in the 1930s, the liquefaction of coal was not economically feasible, so the plant was converted to use hydrogenate creosote oil. During World War II, this plant produced aviation-grade gasoline from creosote.

Emerging Direct Liquefaction Technologies

In the years immediately following World War II, some commercial-scale direct liquefaction continued, but the products could not compete economically with cheap petroleum from the southwestern United States or from the Middle East. During the 1950s, these plants were gradually abandoned. The most significant activity in the United States at that time was a plant built by the Bureau of Mines in Louisiana, Missouri. This plant had a production capacity of 8000 gallons per day of synthetic liquid fuels. The Louisiana facility was closed because a petroleum shortage that would necessitate the production of liquid fuels from coal seemed unlikely.

In the late 1950s and throughout the 1960s, large-scale work in direct liquefaction of coal was virtually nonexistent. Fundamental research on this subject was also neglected during this time. A dramatic reversal of the situation occurred as a result of the energy crisis of the 1970s, beginning with the first oil price shock in 1973. The dependence of the United States and western Europe on imported Middle Eastern oil was demonstrated, as well as the potential for domestic disruption arising from a long-term curtailment of supplies from the Middle East. This situation led to a rapid revitalization of direct liquefaction.

As was the case with gasification, dozens of liquefaction schemes were proposed. Some existed only in the inventor's mind; others were tested in laboratories or small pilot plants; a few even warranted trial in large pilot plants. However, none are currently operating on a commercial scale.

Although many proposed schemes appear to differ greatly, the basic chemical concept of direct liquefaction remains unchanged from Bergius's time: The hydrogen-to-carbon ratio must be increased. Therefore, most recent approaches to direct liquefaction are seeking to improve Bergius's process by the following methods:

- facilitating the transfer of hydrogen to the coal,
- reducing the severity of processing conditions (i.e., operating at lower temperatures and pressures),
- increasing the rates of hydrogenation reactions, or
- increasing the yield of the desired products (e.g., increasing diesel oil at the expense of heavy paraffins).

All of these improvements can potentially reduce the cost of the synthetic fuel produced.

For the processes discussed next, the small-scale technical results and preliminary economic estimates warranted the building and operation of large-scale pilot plants or demonstration plants. In the mid-1980s, falling prices of crude oil have virtually destroyed the likelihood of building a commercial direct-liquefaction plant in the near future. Nevertheless, these processes are the most likely to be commercialized when the need arises.

Exxon Donor Solvent Process

Hydrogen Donation. In the mid-1960s, Exxon began a research program on a liquefaction scheme that eventually became known as the Exxon Donor Solvent process (often referred to simply as the EDS process).

The name implies the key feature of this process: The hydrogen added to the coal mainly derives from hydrogen-donor molecules in the solvent rather than from gaseous hydrogen. After the hydrogen donors have transferred hydrogen to the coal, they are replenished in a separate hydrogenation step. For example, naphthalene is hydrogenated to 1,2,3,4-tetrahydronaphthalene, a compound commonly known as Tetralin. Tetralin, which is present in the solvent mixture used for the liquefaction step, donates hydrogen to the coal to facilitate the breakdown of the macromolecular structure of coal. After it donates hydrogen to the coal, Tetralin converts back to naphthalene. In a separate step, naphthalene reacts with hydrogen gas in the presence of a catalyst to regenerate Tetralin, a step usually called *hydrotreating*.

Pilot-Plant Testing. The liquefaction research was begun in laboratory-scale reactors that used 5 grams of coal per test. As progress continued, the process was tested in small pilot plants that first used 75 pounds of coal per day and then later used 1 ton of coal per day. The culmination of work on the EDS process was the construction and successful operation of a pilot plant that used 250 tons of coal per day in Baytown, Texas. The building and operation of this plant was supported by the U.S. Department of Energy; the Electric Power Research Institute; an international group of industrial sponsors, including Phillips Petroleum Company, ARCO, Ruhrkohle, and Japan Coal Liquefaction Development Company; and Exxon, which provided internal funding.

Operation. In the EDS process, finely ground coal is mixed with the donor solvent, which has been recycled through the process; the slurry is preheated and then added with gaseous hydrogen to the reactor. The temperature and pressure depend on the coal being used; generally, the temperature ranges from 800 to 870 degrees Fahrenheit, and the pressure ranges from 100 to 140 atmospheres. The products from the liquefaction reactor are separated by distillation into light hydrocarbon gases (ranging from methane to propane and methylpropane), a naphtha fraction, a heavy distillate, and a portion that cannot be distilled, which is called the *bottoms*. These operations are illustrated in Figure 11.2.

Product Treatment. The naphtha and heavy distillate fractions are treated by conventional petroleum-refining technology. About 85% of the naphtha is recovered as an excellent grade of gasoline having an octane number of 105. Alternatively, about 50% is recovered as a mixture of benzene, toluene, and xylenes, which is known as BTX. Further processing of the heavy distillate yields fractions comparable to jet fuel and heating oil. A portion of the heavy distillate is hydrotreated and recycled to slurry with fresh coal.

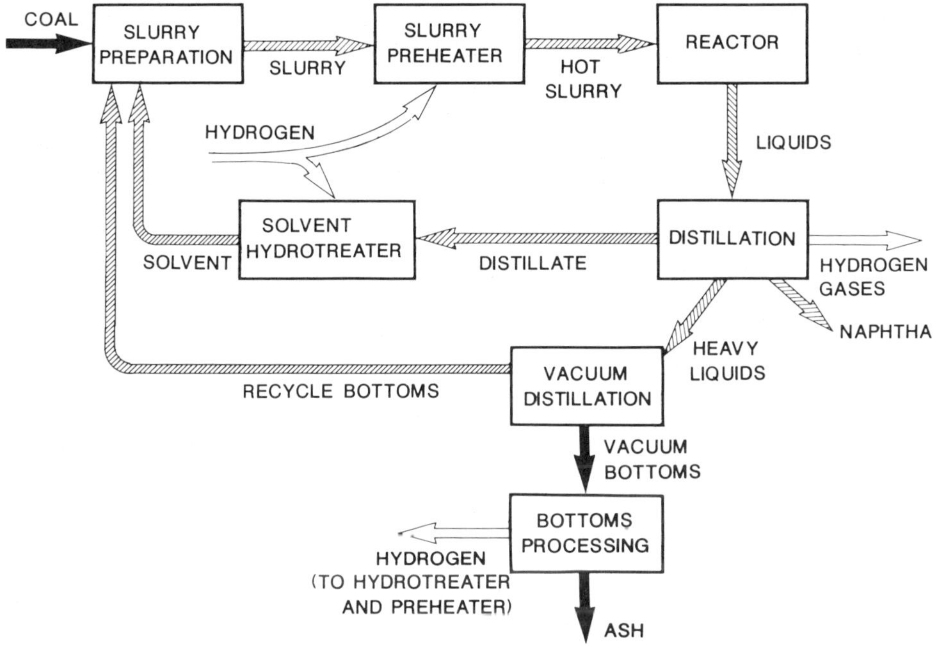

Figure 11.2. The EDS process relies on transferring hydrogen atoms from molecules in the solvent to the coal. The hydrogen-donor molecules in the solvent are regenerated in a separate hydrotreating step. Unlike other approaches to liquefaction, the EDS process does not require the addition of hydrogen to the reactor itself.

The Bottoms. Originally, the bottoms were treated by another Exxon process called *Flexicoking*. The bottoms were converted in the Flexicoker to heavy oil, which could be added to the process streams, and coke. Gasification of the coke in an oxygen-blown gasifier provided a source of hydrogen for the liquefaction reactor; alternatively, an air-blown gasifier produced an inexpensive low-heating-value gas for use as a fuel in the plant.

As research on the EDS and other direct liquefaction processes continued, an alternative use was found for the bottoms. Recycling the bottoms resulted in a dramatic increase in the conversion of the coal to liquid products. For example, 55% of a bituminous coal from Illinois converts to liquids that can be distilled. By operating EDS with bottoms recycle, conversion increases to 75%, and the relative proportion of light (i.e, gasolinelike) liquids in the product increases as well.

Solvent Quality Index. A key feature of the EDS process was a careful matching of the chemistry of the donor solvent with the characteristics of the coal being liquefied. Exxon chemists developed a descriptive parameter called the *solvent quality index,* which is based on painstaking analyses of liquefaction solvents and correlations with process results. The solvent quality index can be adjusted by controlling the reactions in the hydrotreater. One of the coal characteristics that is important in selecting the appropriate solvent quality index is the amount and kinds of oxygen-containing functional groups in the coal structure. Two other features of the EDS process set it apart from many other direct liquefaction processes:

1. No catalyst is added to the liquefaction step.
2. The products are recovered by distillation without needing a filtration step to remove solids from the liquid product.

Technical Success versus Economic Conditions. In terms of the technology, the EDS process was a success. However, the economic realities of falling petroleum prices overtook the technical progress. The Baytown plant, which was once the most visible monument to liquefaction research at Exxon, has now been dismantled.

H–Coal Process

Operation. The H–Coal liquefaction process was developed by Hydrocarbon Research as an outgrowth of previous work on the hydrogenation of petroleum fractions. The reactor used in the H–Coal process is significantly different from that used in the EDS process. Pulverized coal is slurried with a heavy recycled oil. This slurry, together with hydrogen at 170–200 atmospheres, passes through a preheater and then to a reactor at about 850 degrees Fahrenheit. The reactor contains a bed of catalytic particles, usually cobalt molybdate supported on aluminum oxide. The rapid passage of three phases—solid coal, liquid oil, and gaseous hydrogen—keeps the bed in a churning, random motion as if it were boiling. This type of reactor is called an *ebullated-bed reactor*; this term derives from the word *ebullition,* which describes the action of bubbling or boiling.

The key to operating the ebullated-bed reactor is the maintenance of a significant difference in particle size between the coal and the catalyst. The coal is ground to pass through a sieve having 0.15-mm openings, whereas the catalyst particles range from 1.5 to 6.0 mm. In principle, unreacted particles of coal or mineral matter should be swept out of the top of the reactor without carrying any of the expensive catalyst with them.

Product Treatment. The H–Coal products are light hydrocarbon gases, distillable liquids, and bottoms. The liquids are not as readily amenable to treatment by conventional petroleum technology as the EDS products are. Nevertheless, the naphtha is converted to gasoline, and the heavy distillate is converted to jet fuel and heating oil. Treatment of the bottoms in an oxygen-blown gasifier is a source of hydrogen for the process; the bottoms have been successfully tested in a Texaco gasifier. Unlike the EDS process, the H–Coal product stream requires filtration to remove solids (unreacted coal and mineral matter).

Pilot-Plant Testing. The H–Coal process was the subject of extensive research that eventually resulted in the building of a pilot plant that used 600 tons of coal per day in Catlettsburg, Kentucky. The H–Coal pilot plant was still operating in 1985, although continued operation of the plant and commercialization of the process in the near future are questionable.

Solvent Refined Coal Processes

The Solvent Refined Coal liquefaction process referred to as SRC–2 is an outgrowth of an earlier Solvent Refined Coal process tested by Gulf Oil in the 1960s. The earlier process is now known as SRC–1; its goal was the production of a low-sulfur, low-ash, solid fuel. The SRC–1 product was envisioned as a boiler fuel that could replace natural gas or fuel oil and would have significantly reduced problems of ash handling and sulfur oxide emissions in comparison with untreated coal.

Operation. In the SRC–1 process, pulverized coal was mixed with a recycled solvent and hydrogen at 70 atmospheres and heated in a dissolver at 840 degrees Fahrenheit. The product was filtered to remove mineral matter and unreacted coal. After evaporation of the solvent, cooling to lower than 400 degrees Fahrenheit resulted in the formation of a solid having about 0.1% ash, 0.3% sulfur, and a heating value of 16,000 British thermal units per pound (Btu/lb).

In contrast, the goal of the SRC–2 process is the production of synthetic liquid fuels rather than a solid. In the SRC–2 process, the dissolver temperature is higher, up to 870 degrees Fahrenheit, and the hydrogen pressure is also higher, ranging from 100 to 170 atmospheres. Another key modification is recycling some dissolver slurry to slurry fresh coal rather than recycling a portion of the heavy distillate oil. The effect of recycling dissolver slurry is that some unreacted coal is passed back through the dissolver. Thus, the recycling increases the time that coal particles are exposed to dissolver conditions. Furthermore, the reaction conditions are more severe in the SRC–2 dissolver than in the SRC–1 dissolver. The longer time at more severe conditions results in more complete conversion.

Product Treatment. The SRC–2 dissolver slurry is distilled to eliminate the difficult solid–liquid separation step of the SRC–1 process. An oil product that is comparable to bunker C fuel oil (commonly known as No. 6 fuel oil) is obtained. This product, like bunker C fuel oil, is a heavy oil that frequently requires preheating for handling or combustion. (Bunker C fuel oil is prepared by diluting the distillation bottoms with 5–20% distillate.) Other distillation products include naphtha and a middle distillate oil that are converted to gasoline and diesel fuel, respectively. The bottoms, which contain unreacted coal and mineral matter in the SRC–2 process, are treated in an oxygen-blown gasifier to obtain hydrogen for the process. A block flow diagram of the SRC–2 process is shown in Figure 11.3.

The SRC–2 process has the advantage of being relatively simple. The mineral matter in the coal, particularly the iron compounds, appears to catalyze the dissolution step and thus eliminate the need for expensive catalysts. Recovery of products by distillation eliminates filtration or other solids-separation processes. Hydrotreating a donor solvent is also unnecessary.

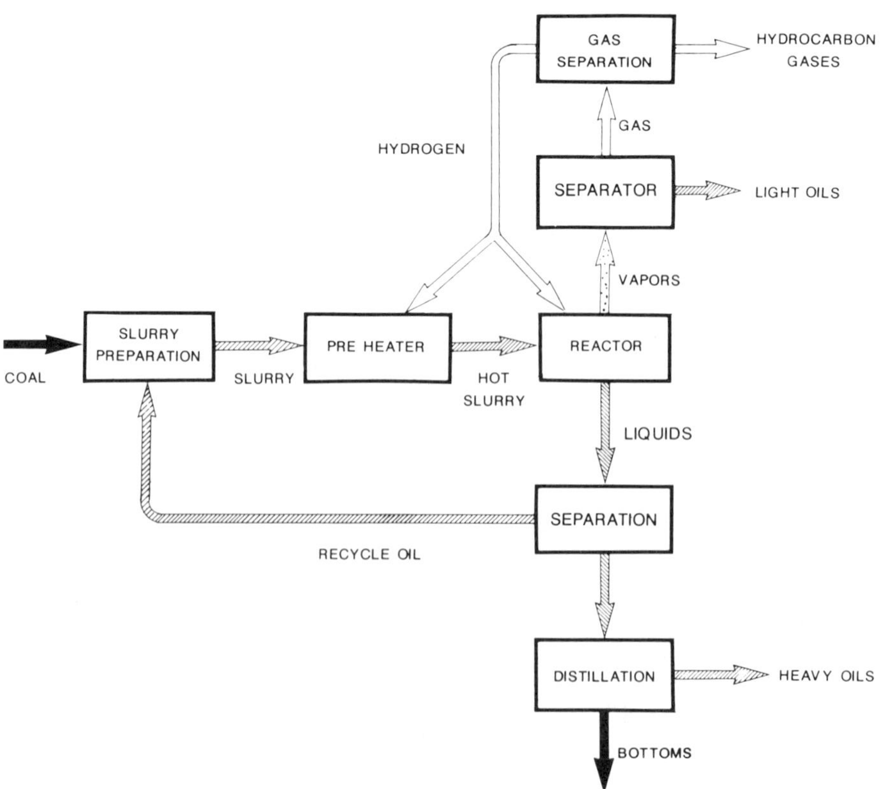

Figure 11.3. The SRC–2 process reacts hydrogen with coal first in the slurry preheater and then in the liquefaction reactor. This process can produce a variety of liquid products.

Pilot-Plant Testing. An SRC–2 pilot plant that used 50 tons of coal per day was built and operated in Fort Lewis, Washington (near Tacoma). Although reasonable technical success was achieved, the waning interest in synthetic liquid fuels led to the closing of the plant. The SRC–1 process, however, has been tested in a pilot plant in Wilsonville, Alabama. This plant was still in operation in 1985.

Conclusions

In technological terms, the production from coal of synthetic liquid fuels that are comparable to petroleum fractions is entirely feasible. Germany and South Africa have shown that a country can meet much of its needs for liquid fuels from coal liquefaction. Without overriding concerns such as war or isolation due to domestic racial policies, economics will be the driving force for the development of a synthetic liquid fuels industry in the future.

If the price of crude oil rises dramatically because of declining production or curtailment of imports, a point will be reached at which the cost of coal-derived liquids is less than the cost of petroleum-derived fuels. This situation will create an incentive for industry to build commercial-scale liquefaction plants. However, predicting if or when this situation will occur is impossible. Nevertheless, continued fundamental research on coal structure and reactivity may lead to ways of hydrogenating coal more rapidly or at lower temperatures and pressures than current processes. These process improvements could result in less expensive reactor vessels or better coal conversion, either of which could reduce the cost of the liquid product.

Notes

[1] Speer, A. *Inside the Third Reich* Macmillan: New York, 1970; Chapter 24.

[2] The octane number is assigned to a fuel by comparing its combustion characteristics in a standard test engine to a mixture of 2,2,4-trimethylpentane (otherwise known as isooctane) and heptane. The octane number is equal to the percent by volume of isooctane in the test blend having the same combustion characteristics as the fuel being evaluated.

Chapter 12

A Look to the Future

The ultimate catastrophe—a nuclear holocaust—would not only destroy our civilization but would likely be the end of life on earth. However, if we assume that we will avoid such a catastrophe, then we must prepare for the continued progress of humanity in the future. In this regard, we must develop energy sources that are inexhaustible and nonpolluting, such as the following:

- sunlight,
- biomass,
- wind, and
- nuclear fusion.

We must also consider that oil and natural gas supplies will become increasingly scarce and expensive.

Whether or not we will soon deplete our supplies of petroleum is unclear. Many energy forecasts have stated unequivocally that we will run out of petroleum in about 10 years. However, some of these forecasts were made in the 1920s, and we have certainly survived the intervening years without the predictions coming true. Some people, particularly Thomas Gold of Cornell University, believe that the supply of petroleum in the earth is far greater than most geologists have imagined. However, unless Gold's belief is proven correct, we must recognize that the amount of petroleum in the earth is finite. Because we are using petroleum at a substantial rate and are not discovering new supplies as frequently as in the past, petroleum products will eventually become less available.

Three options are available for making the transition from an economy based heavily on petroleum to an economy based on nonpolluting and virtually inexhaustible energy.

1. Accept a lifestyle that involves the consumption of much less energy.
2. Increase the use of nuclear energy.
3. Increase the use of coal.

Most likely, people will not accept the inconveniences and disruptions caused by a reduced availability of energy, for example, long lines at gas stations such as those experienced during the oil embargo of the early 1970s. The nuclear option, at least in the United States, will not be selected unless a radical change in public sentiment occurs. Therefore, the transition from our present energy economy, based heavily on petroleum, to an economy that uses renewable or inexhaustible energy supplies is likely to depend on coal. The slogan of the coal industry in the 1970s is an apt one: Coal is America's ace in the hole.

Chapter 1 discussed the needs of a modern industrial society: electricity, transportation fuels, home and industrial heating, reducing agents to make metals from their ores, and chemicals. Chapter 1 also explained how coal is meeting these needs or met them in the past. Unless substantial changes in lifestyles and the standard of living occur, these needs will continue. What are the future prospects for coal?

Electricity from Coal

The main use of coal today is for electric power production. About two-thirds of the coal produced annually is burned in power plants. Coal-fired plants provide about one-half of our total electricity. The pulverized-coal or cyclone boilers are a proven technology. The most straightforward way to increase coal's share of the total energy market is to build more power plants using the current conventional combustion technology. The current technology, however, is not without concerns or problems, including transportation of coal to the plant, sulfur and nitrogen emissions, ash disposal, and efficiency of converting the energy in the coal to electricity. New technologies are emerging to address some of these concerns.

Slurry Combustion

The development that may change coal transportation and some aspects of power-plant design is coal-slurry production. Coal–oil slurries were developed to conserve oil. When these slurries are burned, some of the heat comes from the coal; therefore, less oil is used to produce a given amount of heat than if pure oil was burned. Today, coal–water slurries are the focus of interest. At first, burning something that contains water may seem strange; however, coal–water slurries burn quite well if the coal in the slurry provides more than enough heat to evaporate the water.

Coal slurries offer an alternative to the traditional methods of coal transportation by railroads, trucks, or barges. Fluids are often more convenient to handle than solids because fluids can be pumped through pipelines and delivered for use by opening a valve. Railroads handle about two-thirds of all coal shipped in the United States; thus, coal–slurry pipelines pose a threat to the railroads.

If permission to construct pipelines can be obtained, with or without support of the railroad industry, two technical issues remain. The first issue is the prospect of burning the slurry directly without dewatering to recover solid coal. Developments in burner designs offer this possibility. The second issue is the prospect of combining coal beneficiation with slurry preparation to reduce the problem of ash and sulfur generation in the combustion process. For example, the preparation of lignite slurries reduces the sodium content of the coal and appears to reduce ash deposition on boiler tubes.

Converting existing power plants to the use of slurry fuels is achieved relatively easily. The fuel-handling system and the burners are changed to accommodate the slurry. If the slurry is fired to provide about the same rate of heat generation, and if major changes in pollution-control equipment are not needed, then the rest of the power station requires little modification. Retrofitting to slurry combustion results in simpler transportation of the fuel and more convenient ways of handling the fuel in the plant. In addition, it offers the possibility of switching from oil to coal firing.

Today, power plants based on coal slurries are not much different from conventional pulverized-coal-fired plants. However, new technologies, now in advanced stages of development, may result in power plants that are substantially different in appearance and operation. These technologies are combined-cycle plants and fluidized-bed combustors, both of which are close to appearing in new, large-scale power plants.

Combined-Cycle Plants

The combined-cycle plant uses a gasifier to convert coal to gas. The gas is first burned in a gas turbine; the hot exhaust gases generate steam that drives a steam turbine. The combined-cycle plant promises several improvements in comparison with current technology. The most important improvement is better efficiency in converting coal to electricity. The efficiency of pulverized-coal-fired units is about 36%, whereas that of combined-cycle plants is about 42%. At first this increase may seem to be of little consequence; however, it is a major improvement in performance, which translates into substantially better economics. A second important technical aspect of the combined-cycle plant is the need for a clean gas stream for the gas turbine. This requirement means that many potential pollutants have to be removed early in the process to reduce the need for pollution-control equipment at the end.

Because the combined-cycle plants consist of two separate systems, namely, the gas-turbine cycle and the steam cycle, the plant offers some flexibility in construction. Although a conventional pulverized-coal boiler can be operated somewhat below its designed rating, the range of efficient operation is limited. A utility deciding whether or not to add a pulverized-coal-fired boiler that has a 500-megawatt capacity, for example, is making a choice between 0 or 500 megawatts of new capacity. In contrast, the combined-cycle plant can, in principle, be installed in stages. For example, a first stage can be installed to provide 300 megawatts immediately; later, when the power needs of the utility system increase, a second stage can be added to provide the full 500 megawatts. This option allows the capital investment to be stretched over a period of time.

The most notable example of development of the combined-cycle plant is Southern California Edison's Cool Water station in the Mojave Desert. The heart of the Cool Water facility is a Texaco gasifier, which is an entrained-flow, slurry-fed, slagging gasifier. The Cool Water operation is showing promising results; consequently, combined-cycle plants may come on-line in the 1990s.

Fluidized-Bed Combustors

Fluidized-bed combustors offer a different set of design features than the combined-cycle plants; however, they still represent potential improvements of current technology. The fluidized bed is more tolerant of variations in fuel quality than are other combustor designs. Not only can different types of coals be used, but the unit can burn heavy oil, wood, or refuse-derived fuel. This tolerance gives the utility greater flexibility in arranging fuel supplies and offers freedom from long-term fuel contracts.

Fluidized-bed combustors can operate over wide ranges of output, that is, from the full design capacity to almost idling conditions. This ability also provides the utility with a greater flexibility in operation. The plant operation can be adjusted, at least somewhat, to respond to changes in power demand. The range over which fluidized beds can be operated is greater than the ranges of other combustor designs.

Fluidized beds can also reduce the formation of sulfur oxides and nitrogen oxides. The temperature in the bed is only about 1400–1750 degrees Fahrenheit, which is much lower than bed temperatures of other types of combustors. The bed temperature is lower than the decomposition temperature of many sulfates. Therefore, limestone or dolomite can be added to the bed to react with the sulfur oxides to produce calcium sulfate and magnesium sulfate, which will be retained in the bed. For low-rank coals that produce highly alkaline ash, the ash may serve as an adequate sulfur-capturing agent. The capture of sulfur in the bed shifts the problem from removing sulfur oxides from flue gases to disposing of sulfated-bed material. Nevertheless, this method eliminates the need for a flue-gas desulfurization system, which can account for one-third of the total capital investment of a new power plant that uses pulverized-coal technology. Although the sulfated-bed method involves costs of equipment to add the sulfur-capture agent as well as equipment to recover and dispose of the sulfated material, the prospect of eliminating the desulfurization system is an important economic consideration when a board of directors considers investing in a new plant. The low temperature of a fluidized bed substantially reduces the formation of nitrogen oxides. This situation is particularly true for thermal nitrogen oxides, which arise from the oxidation of nitrogen in combustion air.

Like the combined-cycle plants, fluidized beds are on the verge of being commercialized. A fluidized bed has been operating successfully for several years at Georgetown University. The Northern States Power Company has converted their Black Dog plant in Minneapolis to fluidized-bed operation. Montana–Dakota Utilities is retrofitting a plant in Mandan, North Dakota, for operation in March 1987. The Colorado Ute Electric Association is constructing a new plant at the Nucla station in Montrose County, Colorado, for operation in early 1987. Most likely, fluidized-bed plants will be used increasingly in the 1990s and the 21st century.

Slurry combustion, combined-cycle plants, and fluidized-bed combustors are new technologies for producing electricity from coal; they are likely to be commercialized by the end of the 20th century. In all of these technologies, the heat from the burning coal is converted to mechanical energy in a turbine; the mechanical energy is then converted to electrical energy. These conversions introduce much of the inefficiency in electricity generation. Technologies still in the developmental stage offer the prospect of direct production of electricity without the intervening mechanical step.

Magnetohydrodynamics

Magnetohydrodynamics, commonly known as MHD, is based on the principle that an electrical conductor moving through a magnetic field generates an electric current. In a coal-fired MHD system, coal is burned with preheated oxygen or oxygen-enriched air to generate combustion gases at a temperature of 4900 degrees Fahrenheit. These extremely hot gases pass through a duct in which two opposing walls are electrodes. A strong magnetic field is set up at right angles to the duct. To make the combustion gases electrically conducting, a "seed" of easily ionized material is added.

The seed is usually potassium. The ionized, conducting combustion gases passing through the magnetic field produce the electric current, which flows out through the electrodes.

Although MHD offers a straightforward way of making electricity from the heat energy of burning coal, formidable problems must be solved to further develop MHD. The extreme temperatures of the gas, which are much higher than conventional combustion systems or even gas turbines, make the selection of materials for the duct a major challenge. The problem is compounded by the requirement that two walls be conductors and the other two walls be insulators. Furthermore, in a commercial plant, the duct has to function for long periods of time with minimal maintenance. The potassium used as the seed can cost much more than the coal; consequently, some estimates indicate that 99% recovery of the potassium is needed to achieve economical operation.

Hot gases leaving the MHD duct are used to generate steam for a steam turbine, just as gas-turbine exhaust is used for this purpose in a combined-cycle plant. The heat recovery obtained from adding a steam cycle to the MHD system significantly improves the overall efficiency, which is estimated to range from 45% to 55% for a large-scale MHD plus steam-cycle power plant. A gas plus steam combined-cycle plant can also achieve this efficiency if it incorporates new technologies that are approaching commercialization. The efficiency of the combined-cycle plant and the serious technical problems still facing MHD make it questionable whether or not a commercial MHD plant will ever be built in the United States. Today, most of the research and development of MHD is being done in the Soviet Union, where work has progressed to the pilot-plant scale.

Electrochemical Processes

The next stage in the quest for increased efficiency is to eliminate burning the coal altogether. Obtaining energy from coal means that the carbon in the coal must be converted to carbon dioxide and that the energy liberated in the formation of carbon dioxide must be trapped, at least in part, for conversion to some useful form of work. The most efficient process would be one in which the carbon, oxygen, and carbon dioxide are all at the same temperature (i.e., no energy is wasted in heating the carbon dioxide) in an almost reversible system. (In a truly reversible system, an ideal that can never be attained in practice, properties of the system such as temperature and pressure never vary from those of the surroundings by more than an infinitesimal amount. Consequently, no energy is wasted to heat the products or expand the volume.) One way to achieve a nearly reversible system and keep the reactants and products at the same temperature is to use an electrical cell. Because the main reason for oxidizing coal is to make electricity, the prospect of making electricity directly from coal by using an electrochemical process is intriguing.[1]

The conversion of coal takes place in a special type of electrochemical cell called a *fuel cell.* In common electrochemical cells, such as flashlight batteries, the chemicals that undergo reaction are mixed together in contact with the electrodes; the reaction proceeds when the electrical circuit is completed for electricity generation. In contrast, the reactants in a fuel cell are kept separate and are brought into contact with the electrodes when electricity is needed. Because a fresh supply of reactants is continuously available, the fuel cell never "dies".

Two options exist for fuel cells based on coal: direct cells and indirect cells. *Direct cells* use carbon as the negative electrode. Oxygen undergoes ionization at the positive electrode. Copper and iron have been proposed as positive electrodes, but most studies of direct cells have used much more expensive materials: platinum, silver, or gold. Some direct cells use solutions of sodium hydroxide or sodium hypochlorite as the electrolyte. Much research has focused on using molten salts as electrolytes. Sodium hydroxide, borax, and mixtures of potassium carbonate have been tried. The net chemical reaction in a direct cell is the reaction of carbon with oxygen to form carbon dioxide. Recently, *indirect cells* have attracted the most attention. Instead of using carbon as an electrode, indirect cells use carbon monoxide or carbon monoxide–hydrogen mixtures produced by gasification, reacting these gases with oxygen. Because the carbon of the coal is partially oxidized by conversion to carbon monoxide, an indirect cell recovers only a portion (about 65%) of the energy theoretically produced in the transformation of carbon to carbon dioxide. However, the indirect cell usually generates a higher voltage. A design currently being studied is the molten carbonate fuel cell; this cell uses a carbon monoxide–hydrogen mixture obtained from a gasifier and operates with nickel oxide electrodes in a mixture of molten sodium carbonate and potassium carbonate. Several technical problems remain to be solved, including the electrolytes wetting the electrodes and corrosion of the electrodes. Both direct and indirect cells have been subjects of research for more than half a century; however, more work is needed to make fuel cells commercially successful.

Transportation Fuels from Coal

Steam

Coal can be used as a transportation fuel, or it can be converted to other fuel forms that can be used for transportation, for example, liquid fuels. When the coal-fired steam locomotive was the principal source of motive power on the railways of the world, and the railways were the principal means of moving both people and goods, coal had a significant share of the market for transportation fuels. However, steam locomotives have not been used on major U.S. railroads for more than 25 years (except for occasional special excursion trips for railroad buffs). Nevertheless, considering the steam locomotive as a technology of the past is premature because the traditional steam locomotive has never truly vanished from the world railroad scene. Steam locomotives still provide a major share of the motive power on the railways in China. Furthermore, steam locomotives are still being built in China; approximately one locomotive is produced each day at a cost substantially lower than producing a diesel locomotive of comparable power. A significant factor in the continued use of steam locomotives in China is the availability of enormous reserves of coal and only modest supplies of oil.

If the situation in the United States becomes like that in China (i.e., some coal is still available but not much oil), then resurrecting the old steamers might be considered. Such a move would not necessarily be a major step backward. Some excellent designs developed in the 1940s scarcely had an opportunity to make an impact because of the postwar rush to convert to diesel locomotives in the United States and electric locomotives in Europe. In 1945, the New York Central Railroad began operating the Niagara class locomotives on passenger service between New York and Chicago. The Niagara locomotives generated 6700 horsepower (many modern single-unit diesel locomotives generate 2500 horsepower or less) and covered 928 miles per day; if we consider time for servicing, these locomotives covered 26,000 miles per month. A typical Niagara locomotive traveled more miles in a 24-hour period than many steam locomotives of the time traveled in a week. Niagaras serviced the famous Twentieth Century Limited between New York and Chicago in 16 hours, which is at least 1 hour faster than the best scheduled time today. While the Niagaras were making their mark, Andre Chapelon, possibly the greatest locomotive designer ever, developed a prototype express-train locomotive in France. This locomotive quickly proved to be the most powerful in the world outside of North America. In service on trains between Paris and Lille or Paris and Le Mans, Chapelon's design consistently outperformed electric locomotives. A comparison of coal consumed by the Chapelon locomotive with coal consumed to produce electric power to operate equivalent electric locomotives showed that Chapelon's design had the fuel economy advantage. Use of the Niagaras ended because of disruptions in the coal supply due to postwar strikes of the United Mine Workers and a general deterioration of maintenance and servicing of steam locomotives on the New York Central Railroad. Chapelon's incredibly successful prototype ran afoul of a French bureaucracy committed to electrification of the national railways and never went into production.

Although these designs were good, the return of coal-fired steam locomotives will not have to depend on them or be impelled by an oil shortage. Several firms are now investigating new steam locomotive technology that combines advances in combustion technology with modern locomotive control systems. American Coal Enterprises has developed two locomotive designs that use the reciprocating pistons typical of most steam locomotives. These designs also incorporate a recirculation system to condense and recycle the steam and a system for automatically feeding packages of coal into the firebox. Operating costs for these designs are estimated to be about one-half of those for a diesel locomotive. The National Steam Propulsion Company has designed a locomotive that is essentially a rolling fluidized-bed combustor employing a bed of sand for the actual combustion and a second bed of limestone for capture of sulfur oxides. This system uses high-sulfur coals, which have few markets elsewhere. The relatively low combustion temperatures eliminate the formation of nitrogen oxides and large clinkers of ash.

Coal Slurries

Coal slurries provide another way to use coal as a transportation fuel. The use of coal slurries would not involve a revival of the steam locomotive; instead, coal slurries could substitute for petroleum-based fuel in diesel engines. Railroads spend about $4 billion a year on diesel fuel. Replacement of some of this fuel with less expensive coal slurry could provide tremendous savings. Coal slurries used in diesel engines pose two problems:

1. Water in the slurry causes a longer ignition delay, which can lead to difficulty in keeping the locomotives running at high engine speeds of about 1000 revolutions per minute.
2. Ash produced in combustion can be abrasive and thereby cause accelerated engine wear.

General Electric is one of the companies now working on adapting diesel engines to coal-slurry fuels. This work on large diesel engines for locomotives may possibly be extended to smaller engines in cars and trucks.

Liquid Fuels

Both direct and indirect liquefaction are ways of making liquid fuels from coal. These liquid fuels can replace the fuels made from petroleum that are now used. A country having access to abundant coal reserves can use current technology to obtain all, or at least a substantial portion, of its liquid transportation fuels from coal. This situation has been proven twice, once in Germany during World War II and again in South Africa since the 1950s. Furthermore, the liquefaction processes developed in the 1970s offer improvements over the earlier methods. Nevertheless, unless unusual circumstances are involved, the commercial production of liquid fuels from coal depends on the price of the fuel being competitive with the price of fuel from petroleum.

During the late 1970s, many economic evaluations of coal-liquefaction processes included estimates of the *cross-over point,* which is the price a barrel of crude oil has to reach to be more expensive than a barrel of liquid fuel from coal. Depending on the assumptions of an evaluation as well as who was performing it, the cross-over point ranged between $40 and $80 per barrel. Some people argued that the cross-over point would never be reached because drastic increases in the price of crude oil would cause a general inflation that would continually increase the prices of equipment and components for coal-liquefaction plants. Thus, coal liquids would always be more expensive than oil. This issue died with the steadily falling prices of crude oil in the mid-1980s. In early 1986, the price of a barrel of crude oil dropped almost to $10, a reduction of more than 65% from the high point in the 1970s.

Bernard Acworth, in his book *Back to the Coal Standard*[2], commented as follows:

> *Those of us who for many years have studied the innumerable and conflicting assertions of those interested in the various projects for extracting oil from coal...can feel little surprise that the mind of the general public is in a state of fog on this question, though it has been educated into a belief that if sufficient of its money is employed, or if the colliery companies can be persuaded, by one means or another, to provide the coal and the plant, for Distillation or Hydrogenation, we can obtain all the oil we require from our own coal resources.*

Many observers of the rapid increase and equally rapid decline in coal-liquefaction research from the early 1970s to the early 1980s would likely agree with Acworth, although Acworth wrote his book in 1932! Commercial interest in coal liquefaction comes in cycles, much like predictions of the imminent exhaustion of petroleum reserves. The current consensus is that the production of liquid fuels from coal will not be commercially viable until some time in the 21st century.

The rapid decline in oil prices and the equally rapid decline of interest in coal liquefaction have unfortunately led to a sense of complacency and a tendency to abandon basic research on the molecular structure of coal and the bond-breaking reactions that convert the solid coal to liquid products. The results of such basic research could lead to the development of ways to liquefy coal at less severe temperatures and pressures. These developments would translate into less expensive plant equipment and therefore lower cost products. If we do not continue to make progress in understanding coal liquefaction, even though no commercial plants are being built, then when the day comes that the oil wells really start to run dry, or when some lunatic with a bomb convulses the Middle East, we will be faced with the prospect of once more starting the cycle all over again. In the worst case, to paraphrase an old joke, coal liquefaction is the technology of the future and always will be.

Methanol

An alternative to producing petroleumlike fuels from coal is a variation of the reaction of carbon monoxide with hydrogen. If carbon monoxide and hydrogen (from coal gasification) are reacted at 600–750 degrees Fahrenheit and 300 atmospheres, methanol forms. Methanol has been suggested as a substitute for gasoline and has been tested in engines. A car can be operated successfully on methanol, but some drawbacks occur. On a gallon-for-gallon basis, methanol yields only about one-half of the energy of gasoline. Thus, obtaining the same amount of energy from methanol requires producing, transporting, storing, and handling an amount of methanol that is twice the amount of gasoline that would be used. Engines designed to burn gasoline need to be modified to operate successfully on methanol. Methanol also is a severe poison and can attack the optic nerve, causing blindness. Thus, much care is needed in handling and using large quantities of methanol.

A solution to these problems is the conversion of methanol to gasoline. Mobil has developed a process that uses a zeolite catalyst to help perform this conversion. A key feature of this process is that gasoline-size molecules can pass through the pores of the catalyst; however, larger molecules must stay in the pore system of the catalyst until they break down to gasoline-size molecules. Compared with existing Fischer–Tropsch indirect-liquefaction technology, the Mobil methanol-to-gasoline process has the advantage of producing a much narrower range of products. This narrow range of products from the Mobil process means that only the desired product is forming; therefore, subsequent separation or refining steps are minimized.

Heating Homes and Industry with Coal

In the United States, the use of coal for home heating has almost vanished. In Europe and the developed regions of Asia, coal remains a popular fuel. Producing smokeless-fuel briquettes from coal is an important industry, and the use of these briquettes in densely populated regions has helped reduce the air pollution that would normally result from many coal fires in a small area. In the Third World, the conversion from firewood to coal is now occurring.

Environmental Effects

The introduction of coal or coal briquettes for domestic use has important environmental consequences. As the population in the Third World grows, the needs for fuel for home heating and cooking increase. These increased fuel needs lead to more cutting of forests for firewood. Although wood is usually considered a renewable energy resource, it is not renewable on a seasonal or annual basis as a crop of hay is, for example. Even if a tree cut for wood is replaced immediately with a growing seedling, many years will pass before the new tree will be comparable in size to the freshly cut tree. In addition, a forest is not just a source of wood but rather an ecosystem of plants, animals, and soil. Thus, extensive deforestation causes an ecological catastrophe for animal populations and soil and water quality. Because the trees help to hold the soil in place, cutting forests from hillsides can result in serious mud slides. As the forests are cut back from populated areas, more and more human energy and time are used to gather wood.

The situation now occurring in the Third World has been caused by the same factors that led to the Industrial Revolution in Europe in the 18th century. During the Industrial Revolution, the depletion of wood supplies also forced a switch to coal as an energy source. Scientists cannot yet predict if the development of coal technology in the Third World will cause another Industrial Revolution. If coal improves conditions of poverty-stricken nations to some extent, it will be a positive change.

District Heating

If coal is to be burned in homes, trucks must deliver it and must collect the ash after it is burned. Furthermore, processes are needed to reduce smoke and particulate emissions as well as sulfur oxides and nitrogen oxides. Such processes are less efficient for a multitude of individual furnaces or heaters than for a larger, centralized unit. (In most cases, emission controls are nonexistent on domestic units; in densely populated regions, this lack of emission controls can lead to serious air pollution problems.) However, *district heating* can bridge the gap between individual domestic units and large power stations. This practice is becoming increasingly popular in parts of Europe, but it has not yet been used to a great extent in the United States. District heating uses a centrally located combustion system to generate steam or hot water that is then piped to surrounding homes. District heating is a good application for fluidized-bed combustors because it can take advantage of their ability to be idled or turned down, their low formation of thermal nitrogen oxides, and their capture of sulfur in the bed.

Coal Cartridges

An approach to centralizing the operations of coal preparation and grinding and ash disposal is the use of *coal cartridges,* which are being developed in

Japan. Currently, the coal-cartridge system is scaled for use in industry or district heating rather than for domestic use. Coal cartridges are prepared at a centralized facility where the coal is ground and loaded into containers that can hold about 10 tons of coal; these containers are the *cartridges*. The cartridges are transported by truck to the customers. An entire cartridge can be loaded into a furnace, and the ash is retained in the cartridge. After the coal has burned, the cartridge is collected and taken to a facility for ash disposal.

The coal-cartridge system is a scaled-up version of truck delivery of coal to individual homes and the collection of ashes. It provides the convenience and economy of centralizing coal preparation and ash disposal. Keeping the coal and ash in containers helps control the release of dust or ash to the environment.

Process-Heat and Boiler Industries

Industries can be divided into two categories: process-heat industries and boiler industries. *Process-heat industries* use the heat from burning fuel to cause a change in the material being processed, for example, heating limestone to make lime. The *boiler industries* use the heat to generate some form of power, for example, steam or electricity; the power is then used to perform a process. Most of the chemical industries, including the paper and textile industries, are boiler industries.

The prospects for increased use of coal in the boiler industries are much the same as in the utilities. The scale of the boilers may be different (they are smaller in industries), but otherwise the technological issues are comparable. Coal slurries offer the potential for retrofit applications, particularly if the necessary changes to the burners and the liquid-fuel-handling system are minor. Fluidized-bed combustors have the attractive features of flexibility in the fuel supply and possible elimination of the need for flue-gas desulfurization.

The use of coal for firing kilns in the process-heat industries could increase. *Kilns* are large furnaces used for heating or drying operations. Cement, for example, is produced in kilns by heating a mixture of limestone and clay. A desirable characteristic of kiln fuels is the production of a long flame. Therefore, coals containing high percentages of volatile matter are particularly attractive in this application.

Cogeneration

The concept of *cogeneration* can be applicable to either the process-heat industry or the boiler industry. Cogeneration is used to capture some of the heat that would be wasted if the energy were used for only a single purpose. For example, the steam needed for heating a chemical reactor usually must be at a relatively low temperature. Because combustion generates steam at high temperatures, the energy that could be obtained from the high-temperature steam is lost or wasted when the steam temperature is allowed to drop. However, in a cogeneration system, the high-temperature steam is first sent through a turbine to generate electricity that can be used on site; the low-temperature steam exiting the turbine is then used for the desired process. On the other hand, if a utility needs high-temperature steam to run its turbines, the low-temperature exhaust is ordinarily wasted. However, in a cogeneration system, the low-temperature steam can be used in district heating or supplied to a nearby industry.

Coal Conversion

These points relate mainly to the direct use of coal, that is, burning the coal to liberate the desired heat. An alternative is to convert the coal to a gas or liquid fuel that is then burned. Coal conversion does offer some advantages:

1. Fluids are easier to transport and distribute.
2. Substitute natural gas or petroleumlike derivatives of coal liquefaction can be used in existing gas or oil-fired units without the need for retrofitting to coal firing or installing new coal-fired equipment.
3. The ash content of gas is zero, and that of coal-derived liquids is negligible.

The future use of gas or liquid fuel from coal depends on the same economic factors that surround the use of these fuels in transportation: The technology available today can produce the fuels but not at a cost competitive with petroleum. The job can be done if we are willing to pay the price.

Coal in the Metals Industry

The steel industry today is based on the reduction of iron ore with coke. As long as steel is made by conventional technology, coking coals will be needed to make coke. The key words, however, are "today" and "conventional technology". The steel industry, at least in the United States, is in a period of transition. Blast furnaces, open-hearth furnaces, and Bessemer converters are fading from the scene. The kind of steel industry that will emerge from this transition is not yet known, but most likely the changes will affect the coal industry.

New Technologies

New technologies being investigated include directly reducing iron ore to steel, bypassing pig-iron production in blast furnaces, and developing small steel plants that rely heavily on reusing steel scrap. The period of transition to new technologies is likely to be long. A new blast furnace costs almost $100 million. An investment of this magnitude is an enormous financial strain for many steel companies. Consequently, continuing to operate existing plants as long as possible is strongly favored.

Coke Production

Preferred Ranks of Coal. Another issue affecting the future of coal in the steel industry involves coke. For most future applications of coal, one rank or another may be preferred, but appropriate system designs should allow all ranks of coal to be used, at least in principle. However, coke production requires medium- and low-volatile bituminous coals that have the appropriate swelling properties to produce coke of good quality. A steel industry based on current conventional technology requires possession of, or dependable access to, substantial reserves of coking coals.

Emission Control. Coke production liberates a lot of organic chemicals as byproducts. Without emission-control technology, coke plants have the potential to be serious polluters of air and water. The construction of a new coke plant requires not only coke ovens but also facilities for treating the waste water and for minimizing the escape of gaseous pollutants. A new coke plant, complete with emission-control facilities, costs about $350 million. Because some steel companies have a *book value*[3] of about $500 million, new coke plants are virtually impossible to finance. Therefore, some people feel that America's last coke plant has already been built.

Alternatives. The restrictions on the rank of coal used for coking and the cost of coke plants provide the impetus for seeking alternatives. In the past, research has occasionally been directed to studying the possibility of making coke from other ranks of coal or to extending the ranges of coals that can be blended to produce coke. So far, none of these projects has been unequivocally successful. Another approach has been to investigate injecting the fuel with the air blast through the *tuyères* (nozzles through which air is forced into a blast furnace). Low-rank, noncoking coals are attractive fuels for this application. The rapid evaporation of the moisture in the low-rank coals caused by the great heat inside the furnace blows the coal particles apart, much like the explosion of a popcorn kernel popping. The popcorn effect generates ample surface area of the blast-furnace coke on which reactions can occur.

Reducing Agents. Although the types of technology will change as we progress into the 21st century, carbon in one form or another almost certainly will remain the cheapest reducing agent. Most likely, coal will remain the cheapest and most accessible form of carbon. It will continue to be used, either by conversion to coke or another form of carbon or without conversion to produce metals that are needed in large tonnages at relatively low cost.

Coal represents more than just a source of carbon for metallurgical operations. Gasification converts coal to carbon monoxide and hydrogen; both are useful reducing agents. Nickel is made by reducing its oxide with water gas. The nickel can be purified by reacting it with carbon monoxide to make nickel carbonyl; nickel carbonyl vaporizes from impurities and is decomposed to pure nickel by heating. Both tungsten and molybdenum are made by reducing their oxides with hydrogen. These metals serve as valuable alloying agents for specialty steels; they produce steels that have good heat resistance, high fatigue strength, and desirable performance as tool and die steels. Germanium is also made by hydrogen reduction of its oxide. Refining the germanium gives a high-purity material used by the semiconductor industry.

Chemicals from Coal

In the early 20th century, the byproducts of coke manufacturing were the foundation of the organic chemical industry. As the demand for organic chemicals, for example, fertilizers, plastics, and synthetic fabrics, continued to grow, the coke industry could not keep up with the demand. At one time, 95% of the phenol consumed in the United States came from the coke industry. The tonnage of phenol used remained constant for many years; however, the chemical industry grew at such a rate that the phenol made as a coke byproduct eventually represented only about 5% of the total consumption. Today, most of the organic chemical industry relies on petroleum for raw materials.

Because a large portion of the chemical industry depends on the use of petroleum, coal is likely to replace petroleum as a fuel when petroleum supplies decline. Then, the chemical industry can continue to use petroleum as a source of chemicals. However, when petroleum supplies decline even more, coal will also replace petroleum as a source of chemicals. Nevertheless, the new age of chemicals from coal will not be a return to the old technology that served so well more than half a century ago. The coke industry simply cannot provide enough byproducts to meet current needs for chemicals. Instead, the new coal chemical industry will be based on new processes.

Coprocessing

The technology is already available for producing substitutes for natural gas and petroleum from coal. As discussed earlier, building a full-scale plant for coal gasification or coal liquefaction has no economic incentives today. However, *coprocessing* provides a way for coal to be used in the liquid fuel and petrochemical industries. Coprocessing combines coal and petroleum processing in the same plant. One option is to use coal-derived liquids as a supplement to the petroleum feedstock in a refinery. Another alternative is to use the heavy residue from petroleum distillation operations as a solvent or vehicle to slurry coal for feed to a coal-liquefaction reactor. These ideas offer some promise but are not without potential problems. The aliphatic compounds in petroleum are not good solvents for coal. A more fundamental concern is that the industrial organic chemistry based on petroleum is largely the chemistry of aliphatic compounds, whereas coal chemistry is largely the chemistry of aromatic compounds. Therefore, conversion to a new chemical industry based on coal may not be as straightforward as using liquids from coal to feed existing chemical plants.

Synthesis Gas

In today's organic chemical industry, the main chemicals are synthesis gas (a mixture of carbon monoxide and hydrogen), ethylene, propylene, 1,3-butadiene, benzene, toluene, xylenes, and phenol. The list of primaries from coal-tar chemistry was somewhat different: benzene, naphthalene, anthracene, toluene, xylenes, phenol, cresols, and phenanthrene. Also, coke-byproduct chemistry involved working with mixtures of dozens of compounds, most of which were produced in quantities ranging from modest to minuscule. Current chemical technology focuses on a much narrower range of materials that are produced in large amounts.

Synthesis gas can be converted to chemicals by well-established technology. Today, synthesis gas is made by reacting methane or naphtha, which contains mostly pentanes and hexanes, with steam. Once the synthesis gas has been made, its subsequent conversion to chemical products does not depend on the original source of the synthesis gas. Therefore, synthesis gas can be made from coal and then used in exactly the same manner as synthesis gas from methane or naphtha. However, coal is not a desirable starting material because a plant based on coal requires facilities for unloading, handling, and grinding the coal. Also, the raw gas from the gasifiers may have to be cleaned before use in the synthesis reactors. Furthermore, the ratio of hydrogen to carbon in coal (about 0.8) is much lower than the ratio in naphtha (2) or natural gas (4); consequently, the hydrogen to carbon monoxide ratio in the synthesis gas from coal may need to be adjusted to meet the requirements of the synthesis steps.

Despite the disadvantages of using coal instead of methane or naphtha, the conversion of synthesis gas to chemicals via Fischer–Tropsch technology is the route to a variety of products that are useful themselves or that can be converted to other materials. Depending on the choice of temperature, pressure, and catalyst, synthesis gas can be converted to substitute natural gas, saturated aliphatic hydrocarbons (paraffins), unsaturated hydrocarbons (olefins), or aromatic compounds. The oxo process, for example, makes alcohols by reacting olefins with synthesis gas. Both synthesis gas and olefins can be made by coal gasification. Because the technology for converting synthesis gas to chemicals is well-known, the initial return of coal to the chemical industry may be via synthesis gas.

Hydrocracking

Products of the liquefaction, pyrolysis, or solvent extraction of coal can be reacted with hydrogen to break apart the aromatic molecules. This process, called *hydrocracking,* produces naphtha and liquefied petroleum gas (LPG), which contains mostly propane and butanes. In the past, the goal of coke production was the manufacture of a solid product; the gases or liquids produced were byproducts. In new pyrolysis processes for chemical production, the objective is to maximize the liquid yield and find some uses for the solid char byproduct. The char can be burned to help meet heat or steam requirements in the plant, gasified to generate synthesis gas or hydrogen, or possibly even burned in a power plant.

Supercritical-Fluid Extraction

A new approach to the extraction of coal uses *supercritical fluids,* which are materials used at temperatures and pressures higher than their critical points; a *critical point* is the temperature and pressure at which the gas and liquid phases of a substance in equilibrium with each other can no longer exist separately. (The process is also called supercritical-gas extraction or supercritical-solvent extraction.) As a new way of processing coal, supercritical-fluid extraction offers some intriguing possibilities. Because the supercritical fluid has no surface tension, it can penetrate the entire internal pore system of coal to the point at which access to the pores is limited only by the physical bulk of the molecules of the fluid. At the same time, the supercritical fluid can act much like a true liquid in its ability to dissolve components of the coal and transport them out of the system. If the solvent ability of a supercritical fluid can be matched to a particular class of compounds in the coal, then the compounds of that class can be selectively extracted. Selective extraction of compounds of particular interest eliminates the need to separate complex mixtures to obtain these compounds. Low-molecular-weight compounds extracted by a supercritical fluid can be treated with hydrogen to remove oxygen, nitrogen, and sulfur; this treatment produces a clean, distillable product.

As a process for making chemicals from coal in the future, supercritical-fluid extraction offers a number of attractive features, including the possibility of using cheap fluids, such as water or carbon dioxide, or tailoring the fluid for selective extraction. The leftover char is fairly reactive and is a good candidate for gasification; it can even serve as a low grade of activated carbon for waste-water treatment. These attractive features need to be balanced against the operation of the process at elevated temperatures, which adds an operating cost for heating, and at elevated pressures, which adds an increased capital investment for the reaction vessels needed to withstand the pressures.

Acetylene

An alternative route to making chemicals from coal is based on the chemistry of acetylene. At one time, acetylene was used as a starting point for the synthesis of many organic compounds, but today it has largely been displaced by ethylene. Acetylene was made by first reacting coal with calcium oxide to produce calcium carbide. The subsequent reaction of calcium carbide with water produced acetylene. If a significant increase in the price of ethylene should occur, perhaps because of a drastic reduction in the availability of petroleum, acetylene could again become an attractive option as a chemical feedstock. If this situation occurs, however, a return to the synthesis using calcium carbide is unlikely. Calcium carbide was made in electric arc furnaces, which are difficult to scale up and which must be located near sources of large amounts of cheap electricity. Instead, the synthesis of acetylene will probably involve the rapid pyrolysis of coal in the presence of hydrogen. If the reaction is run at temperatures exceeding 3600 degrees Fahrenheit and the particles are heated for only one-thousandth of a second before being quenched, acetylene is a major product. One type of reactor that achieves such extreme temperatures for very short times is a rocket engine. Research is now being done to adapt rocket-engine technology to fast coal pyrolysis for acetylene production; this approach is certainly novel for making chemicals from coal.

Coalfineries

Besides new processes being developed for making chemicals from coal, new facilities will have to be built. The combination of several types of processes into a single plant is likely. These plants are called *coalfineries* or *coalplexes*. Depending on the possibilities for marketing the products, a coalfinery could incorporate any of the processes for producing synthetic fuels or chemicals discussed in the past several chapters.

In a simple case, supercritical-fluid extraction of the coal could be used to obtain an extract that, when treated with hydrogen, would yield chemicals. Some of the char remaining from the extraction process could be gasified to make the hydrogen needed to treat the extract. The rest of the char could be burned in a fluidized-bed combustor to make some of the process heat or steam needed at the plant.

A more ambitious approach to the coalfinery concept would be the power–oil–gas–other (POGO) process. The POGO process would combine gasification, pyrolysis, and Solvent Refined Coal (SRC) liquefaction operations. The gasifier would provide gas to a combined-cycle system for power generation. The pyrolysis operation would produce substitute natural gas, LPG, and a gasolinelike liquid. Liquefaction would yield gasoline and fuel oil. Any undistillable fractions from the liquefaction process could be converted to a coke by a high-temperature treatment.

A coalfinery set up principally for production of chemicals would probably rely on liquefaction and gasification feeding Fischer–Tropsch reactors. Fuel products from such a plant would range from substitute natural gas, LPG, gasoline, and diesel to heavy fuel oil. The likely chemical products are olefins, namely, ethylene, propylene, and butenes; aromatic compounds, namely, benzene, toluene, naphthalene, and xylenes; and a variety of alcohols from the oxo process.

A coalfinery could be a versatile coal-processing plant producing a range of fuels, chemicals, and electric power; these products could be tailored to a variety of potential markets. Much of the technology for a coalfinery is already in existence, such as Fischer–Tropsch reactors, or is in an advanced stage of development, such as the SRC process. The main problem to be overcome is finding someone willing to invest in a coalfinery. In today's economy, in which a barrel of oil costs between $18 and $19, finding investors willing to support a plant based on a single technology, that is, a gasification plant or a liquefaction plant, is virtually impossible. Similarly, finding investors for a complex plant combining two or three new technologies is also unlikely, at least in the near future.

How Long Will the Coal Last?

When petroleum and natural gas supplies begin to decline, a new energy economy based on coal can be phased into operation. Coal can provide the electric power, heat, fuels, chemicals, and metals needed to sustain a modern industrial society. For the most part, coal will provide these needs by relying on emerging technologies that can be used in the 21st century with greater efficiency and much less pollution than the old technologies.

In the far future, that is, the 22nd century and beyond, the basis of the energy economy will have to be virtually inexhaustible or truly renewable energy sources. The issue of concern is whether or not our coal supplies will last this long. Unfortunately, we do not know how long our supplies of coal will last.

Factors Affecting Coal Supplies

Amount of Coal Available. The first uncertainty is the amount of coal actually available. Estimates of the amount of coal in a deposit are often based on measurements of coal outcrops or are determined from drill cores spaced a considerable distance apart. These estimates assume that the deposit is continuous between the outcrops or holes and that the thickness of the deposit is approximately constant. Better characterization data may reveal that such assumptions are not justified. Consequently, estimates of the amounts of coal in various deposits are being continually revised upward or downward on the basis of new information. Furthermore, the earth may contain more coal deposits than those currently known or assumed to exist. The oil price increases of the 1970s spurred a tremendous boom in exploration for new oil deposits. Similarly, when increased coal use strains existing coal reserves, exploration for coal deposits will be expanded.

Amount of Extractable Coal. A second uncertainty is how much coal can actually be extracted. Current mining technology has limitations of depth, seam thickness, and, for underground mines, strength of the roof rock. New developments in mining technology may transcend these limitations and allow the extraction of a greater fraction of the available resources. For example, robot miners controlled from the surface might be able to function under conditions that are intolerable to human workers.

Quality of Coal. A third uncertainty is the quality of coal that we are willing to tolerate or able to use. Today, some coals are unacceptable for use because of their high sulfur, ash, or moisture contents or because of their low heating values. If we become desperate for energy, a high sulfur content may be tolerated. Alternatively, new developments in coal beneficiation could provide ways of cleaning a high-sulfur coal to give acceptable performance. We may someday use the lignites of the Balkans, which contain more than 10% organic sulfur.

Our tolerance of the disruption caused by mining operations and the transportation costs of coal must also be considered. For example, Antarctica contains coal deposits. Will we someday accept commercial mining operations on that continent? Will we be willing to pay the costs of moving that coal to Argentina or Chile?

Rate of Coal Use. A fourth uncertainty is the actual rate at which coal will be used. Coal consumption was remarkably steady for more than half a century, from before World War I to the oil price shocks of the 1970s; consumption fluctuated only occasionally during wars and depressions. Coal use has increased since the 1970s. The question is whether it will level off again for another long period or whether it will continue to rise and, if so, at what rate. Dealing with this uncertainty requires making some fundamental assumptions about the future strength of the economy and future energy needs and costs.

Predictions

Although several uncertainties affect forecasts, some conservative estimates of how long we can rely on coal can be made. The United States has about 204 billion tons of coal that can be economically recovered by existing mining technology. Let us assume that no more coal will be discovered and that no new technologies will be developed to allow for the mining of the hundreds of billions of tons of coal that cannot be extracted by current methods. In 1982, the United States produced (and presumably consumed) 779,300,000 tons of coal. Suppose further that as petroleum and natural gas supplies decline, coal use will increase at a rate of 5% per year starting, for the sake of argument, in 1982. (An increase of 5% per year may not seem like much, but it doubles the original amount every 14 years. In addition, when the plateau in coal usage for half a century is considered, a steady increase of 5% per year is probably beyond the wildest fantasies of the coal industry.) Under these circumstances, our coal reserves will last 114 years, or almost to the year 2100. The assumptions of no new coal discoveries and no advances in mining are quite pessimistic. Whether or not a 5% annual growth rate can be sustained for a century is uncertain. Nevertheless, we are probably safe in saying that we can depend on coal for at least 100 years.

Now the question becomes "What will 100 years buy us?" The answer comes from our experience with other energy sources. Michael Faraday discovered the principle of the dynamo in 1831. About 60 years later, in the late Victorian era, the use of electric power began to have a significant effect on industry and commerce. Edwin Drake drilled the first oil well in the United States in 1859. A petroleum-based energy economy, which involved the extensive use of automobiles as well as the beginning of commercial aviation and diesel locomotives on railroads, emerged in the 1920s. Again, a lapse of about 60 years occurred between the beginning of the technology and its widespread effects. Today, the energy technologies of the long-term future—solar, geothermal, fusion, and wind energy—are in early stages of development. If once again a 60-year lag period occurs between the first stirrings of a new energy source and its extended use, we can probably expect significant commercial development of these energy technologies in the mid-21st century. Until then, we have ample supplies of coal to meet our energy needs.

What Should We Be Doing Now?

When this book was written in early 1986, the price of oil was dropping steadily. The Organization of Petroleum Exporting Countries appears to be in disarray. The world seems to be awash in oil that is less than half the price it was in 1979, although it is not as cheap as it was in 1970. Some analysts have correlated the rapid decline in oil prices to the equally rapid rise in the Dow–Jones industrial stock average. The fervor for coal development that existed in 1976 has just about vanished in 1986. Many of the corporations, including a number of large oil companies, that were vigorously active in coal research or development in the early 1980s have abandoned the field. The Office of Management and Budget is proposing to slash the research appropriations for fossil energy research in the Department of Energy by about 50%.

Unfortunately, this situation has the aura of *Mad* magazine's Alfred E. Newman's remark, "What, me worry?" Almost certainly the day will come when we need to rely more heavily on coal than we do today. The time to prepare for our increased dependence on coal is now, not the day we once again wake up to block-long lines at gas stations or rationed supplies of home-heating oil. This leads to the question, "What should we be doing now?"

Determination of Coal Supplies and Coal Quality

Our first task is to determine how much coal we really have as well as its quality. This determination requires a program of coal exploration and more detailed mapping and characterization of the existing coal deposits. This work is slow and painstaking. Sometimes data must be obtained over many years before a clear picture emerges. The United States seems to be lagging behind some other countries in this endeavor. For example, the massive brown coal deposit in the Latrobe Valley in Australia has been extensively mapped and characterized, particularly by the State (of Victoria) Electricity Commission. Planners already know the quality of coal that will be mined 5 years from now. In contrast to this approach, some U.S. coal companies drill only the minimum number of cores required to sell the coal.

We are not certain when we will need to rely extensively on our coal resources in response to the next energy crisis. Yet, whenever that day comes, we will need to know the extent and variability of coal seams, their depth and pitch, and the characteristics of the coal they contain. With this knowledge we will be able to plan how to meet our needs for electricity, fuels, heat, chemicals, and metals. If we do not have this knowledge, then the next energy crisis will create a situation similar to what happened in the 1970s.

Determination of Safe Mining Practices and Optimal Land Reclamation Methods

Our second task is to determine how to mine coal safely and with minimal long-term damage to the environment. The safety of underground mines and the working conditions of miners have greatly improved since the early days of mining. Nevertheless, miners are still maimed or killed despite these improvements. Research is still needed on improved mining techniques and equipment to reduce underground hazards even more. Similarly, the reclamation of strip mines has helped to ameliorate the devastation left by unprincipled mining companies in the past. Yet, we must realize that some short-term disruption of the environment is unavoidable in strip mining. We need to learn more about how to minimize this disruption and how best to reclaim the mined lands after the coal is removed.

Development of Improved Processes for Burning Coal

In the near future, combustion will be the predominant technology for coal use. Therefore, our third task is to develop improved processes for burning coal to increase the efficiency of converting the energy in coal to electrical energy or process heat. Also, because the burning of coal produces sulfur oxides, nitrogen oxides, and ash, we need to continue to seek more efficient and economical ways of dealing with these materials so that their effect on the environment is minimal.

Development of New Processing Concepts

As time progresses, we will increasingly depend on coal for fuels and chemicals. Therefore, our fourth task is to probe the chemistry of coal to understand better how to dismantle the molecular structure into small, usable molecules. Such investigation will lead to new processing concepts. Some of these concepts will need to be tested in pioneering plants at almost commercial scale. Beginning in mid-1985, the Great Plains Gasification Associates plant in Beulah, North Dakota, began getting the attention of the news media because of a series of squabbles among the original investors, the Department of Energy, and the customers. These arguments involved such issues as the repayment of the construction loans, the actual ownership of the plant, and the honoring of the price contracts. The committee meetings, hearings, and legal actions that ensued ignored the fact that substitute natural gas is not the most important product of the Great Plains plant. The most important product is the engineering know-how that was gained in designing, constructing, and operating a large-scale synthetic fuels plant. The relatively small contribution Great Plains makes to the total national supply of natural gas can be replaced. Furthermore, the replacement gas will probably cost less than the gas from Great Plains. However, the contribution that Great Plains makes to our knowledge of synthetic fuels production cannot be replaced.

The Ultimate Challenge

Now we ask ourselves, "From where will the breakthroughs in coal technology come?" Richard Wheeler, a giant among coal researchers in the early 20th century, stated the following:

> *When I am asked what particular research on coal would be of most practical value to those who have to sell it, equally with those who have to use it, I have no hesitation in saying "Research on the composition of coal."*

Dirk van Krevelen, an equally illustrious coal scientist who was most active in the 1950s, had similar thoughts:

> *Difficulties of a major nature in technical practice nearly always originate from unexpected peculiarities in the compositions of the materials used. This may be the composition on a macroscopic, microscopic, or even molecular level. These unexpected and mostly unobtrusive peculiarities can only be evaluated by the coal scientist.*

In other words, the key to making better use of coal now and to finding new ways of using it in the future lies in improving our understanding of the composition and properties of coal and of the fundamental chemistry of coal reactivity.

Gaining this improved understanding is not an easy job. In the realm of inanimate matter, coal may be the material most difficult to understand. The term "coal" does not even refer to one single material. Instead, coal is several materials ranging from the wet, crumbly, brown coal to the "black diamond", namely, anthracite. Coal is a heterogeneous material and does not have a defined molecular structure like polymers and biological macromolecules do, for example. In addition, coal begins to oxidize immediately upon exposure to air, it does not dissolve completely in any solvent, and it decomposes when heated. Furthermore, the composition and properties of coal may vary over just a short distance within a single mine. To make matters worse, some fundamental properties such as moisture content and porosity are not well-understood and remain subjects of controversy among coal scientists.

Despite these difficulties of coal research, progress toward the better use of coal will come from laboratories in which the jigsaw puzzle of coal is being solved piece by piece and step by step. Coal may indeed be a partially soluble, heterogeneous, highly variable, and poorly characterized material. Nevertheless, scientists who have devoted their careers to the study of coal have found the challenges stimulating and rewarding. As Daniel Boorstin said in *The Discoverers*[4], "The most promising words ever written on the maps of human knowledge are terra incognita—unknown territory."

Notes

[1] The notion of making electricity directly from coal has intrigued scientists for a long time. The prospect is discussed in *Thermodynamics* by G. N. Lewis and M. Randall, which was first published by McGraw-Hill in 1923.

[2] Acworth, B. *Back to the Coal Standard*; Eyre and Spottiswoods: London, 1932.

[3] The *book value* of a corporation is its net assets, calculated by subtracting its debts from its total assets; in other words, it is the actual value of the corporation.

[4] Boorstin, D. *The Discoverers*; Random House: New York, 1983.

Appendix

Suggestions for Further Reading

Chapter 1

The most comprehensive source of information on coal is *Chemistry of Coal Utilization* (Wiley: New York). The first two volumes, edited by Homer Lowry, appeared in 1945. A supplemental volume, also edited by Lowry, was published in 1963. A second supplemental volume, edited by Martin Elliott, came out in 1981. These four volumes are the standard place to start to learn about a particular coal topic (except for mining, which is not covered).

Three other books are generally useful: *An Introduction to Coal Technology* by Norbert Berkowitz (Academic: New York, 1979) provides a good survey of most topics on coal except mining. Some of the sections on synthetic fuels research and development are out-of-date, but the time lags involved in writing and publishing books mean that probably no book can stay current for long in the changing world of synthetic fuels technology. Berkowitz assumes more technical background on the part of his readers than I have done in writing this book; I have used Berkowitz's book as a text in a survey course for seniors and graduate students in chemical engineering. *Coal* by D. W. van Krevelen (Elsevier: Amsterdam, 1981) is now out-of-date in parts; nevertheless, it provides a solid introduction to coal science and an excellent summary of what was known about coal prior to the application of modern scientific instruments in the 1970s. *Coal Utilization: Technology, Economics, and Policy* by L. Grangier and J. Gibson (Graham & Trotman: London, 1981) contains a discussion of the economics and policy decisions affecting coal use, moreso than the other books listed here. The book is written mainly from the perspective of Great Britain and western Europe, but U.S. readers will find much useful information here.

The reader seeking information on any of the specific topics treated in Chapters 2–12 should also consult the first four books discussed in this section. Another excellent reference, which may require digging for information at times, is the *Encyclopaedia Britannica* (Encyclopaedia Britannica: Chicago).

Chapter 2

An excellent introduction to the science and practice of geology is available in John McPhee's *Basin and Range* (Farrar, Straus & Giroux: New York, 1981). McPhee's later book, *In Suspect Terrain* (Farrar, Straus & Giroux: New York, 1983), is also worthwhile reading. A readable history of the earth is *Down to Earth* by Carey Croneis and William Krumbein (University of Chicago Press: Chicago, 1961). A brief overview of coal formation, which contains some useful suggestions on how amateur scientists can study coal balls, can be found in *Fossils for Amateurs* by Russell MacFall and Jay Wollin (Van Nostrand Reinhold: New York, 1972). Many fine introductory textbooks on geology and botany also supplement the material in this chapter.

An outstanding discussion of the organic chemistry of coal formation and geochemistry is Peter Given's "An Essay on the Organic Geochemistry of Coal", which appears in *Coal Science: Volume 3* by Gorbaty, Larsen, and Wender (Academic: New York, 1984). The analogous inorganic chemistry is treated in *Geochemistry of Coal* by Vladimir Bouska (Elsevier: Amsterdam, 1981). Many of the examples in this book pertain to coal and peat deposits in central Europe, which are unlikely to be familiar to U.S. readers. Nevertheless, the book provides a useful discussion of some aspects of the inorganic geochemistry of coal.

Chapter 3

Among the best sources of information on coal resources, which do not require wading through reams of statistics, are articles on the relevant countries in the *Encyclopaedia Britannica* (Encyclopaedia Britannica: Chicago). A vast literature exists on the war between Germany and the Soviet Union. Two books that discuss the effects of Russian coal resources on German war strategy are *The Russo–German War 1941–45* by Albert Seaton (Praeger: New York, 1971) and Bryan Fugate's *Operation Barbarossa* (Presidio: Novato, CA, 1984). The low-rank coals of the United States, which generally have not been as well-documented as the bituminous coals, are discussed extensively in Volume 2 of the *Low-Rank Coal Study,* a six-volume report published by the U.S. Department of Energy in 1981 (Government Printing Office: Washington, DC, 1981). The problems caused by the Saar and

Ruhr coals in developing the peace treaty after World War I are mentioned in *The Kings Depart* by Richard Watt (Simon and Schuster: New York, 1968). Some glimpses of the importance of coal in World War II can be found in Albert Speer's memoir, *Inside the Third Reich* (Macmillan: New York, 1970). Australian coal deposits are discussed concisely in *Energy for Australia* by Arthur Corbett (Penguin: Sydney, 1976).

Chapter 4

Many of the topics treated in this chapter are covered in three books published by the American Chemical Society. Two of these are *Organic Chemistry of Coal,* edited by John W. Larsen (American Chemical Society: Washington, DC, 1978), and *New Approaches in Coal Chemistry,* edited by Bernard D. Blaustein, Bradley C. Bockrath, and Sidney Friedman (American Chemical Society: Washington, DC, 1981). Both books were published as part of the ACS SYMPOSIUM SERIES. The third book is *Coal Structure,* edited by Martin L. Gorbaty and K. Ouchi and published in the ADVANCES IN CHEMISTRY SERIES (American Chemical Society: Washington, DC, 1981). Other useful books are *Coal Structure,* edited by Robert Meyers (Academic: New York, 1982), and the series of volumes entitled *Coal Science,* published by Academic Press approximately every 2 years starting in the early 1980s and edited by Martin L. Gorbaty, John W. Larsen, and Irving Wender.

Chapter 5

Most likely because of the horrific conditions associated with underground mining and the working and living conditions of miners, an enormous number of books have been published on mining. An interesting book that traces the interrelationship between mine accidents and the growth of the labor movement is *The Breaker Whistle Blows* by Ellis Roberts (Anthracite Museum Press: Scranton, PA, 1984). During the 1930s, the plight of miners caught the attention of several leading writers. Edmund Wilson described the activities on behalf of striking miners in selections from his notebooks collected by Leon Edel and published as *The Thirties* (Farrar, Straus & Giroux: New York, 1980); Wilson also wrote an article entitled "Frank Keeney's Coal Diggers", which has been reprinted in *The Portable Edmund Wilson,* edited by Lewis Dabney (Penguin: New York, 1983). Wilson and his colleagues were involved with miners in Kentucky in 1932. Five years later, the equally serious conditions on the other side of the Atlantic were described by George Orwell in his book, *The Road to Wigan Pier* (Harcourt Brace Jovanovich: New York, 1958). The incredible story of the men who survived underground for more than a week following the Nova Scotia disaster is told in *Miracle at Springhill* by Leonard Lerner (Holt, Rinehart and Winston: New York, 1960).

The brief but violent and colorful history of the Molly Maguires is told in several books, including *The Mollie Maguires and the Detectives* by Allan Pinkerton (head of the detective agency) (Dover: New York, 1973), *The Molly Maguires* by Wayne Broehl (Vintage: New York, 1964), and *Lament for the Molly Maguires* by Arthur Lewis (Pocket Books: New York, 1969). The Molly Maguires had an adversary even more powerful than the Pinkertons, namely, Sherlock Holmes. Their encounter with Holmes is part of the story of "The Valley of Fear" (*The Complete Sherlock Holmes*; Doubleday & Co.: Garden City, NY, 1930), which is probably the least-known of the full-length Holmes adventures. It is available in various collections of the Holmes canon. The story of the Molly Maguires and their undoing by the Pinkerton agent James McFarlan was the subject of a 1970 film, *The Molly Maguires,* that featured Sean Connery and Richard Harris.

Probably the best fictional treatment of coal mining is Richard Llewellyn's *How Green Was My Valley,* which is available in paperback (Dell: New York, 1967) and has been made into a movie for television. A novel that has a coal-mining theme is John Yount's *Hardcastle* (St. Martin's: New York, 1980). The fictional town of Gibbsville, the setting of some of John O'Hara's novels and short stories, is modeled after the real town of Pottsville, Pennsylvania. Many of the Gibbsville stories convey the flavor of life in the coal fields.

Mining has been the subject of folk songs. Besides "Sixteen Tons" and "The Ballad of Spring Hill", which were mentioned in the text, a song that sometimes shows up in recorded folk anthologies is "The Bells of Rhymney" by Jan Davis. A recent song is "We Work the Black Seam Together", written and recorded by Sting.

Chapter 6

Coal beneficiation is treated in *Fundamentals of Coal Beneficiation and Utilization* by Shirley Tsai (Elsevier: Amsterdam, 1982). Some good examples of the practical side of coal beneficiation are provided in the collection edited by Paul Merritt, *Coal Age Operating Handbook of Coal Preparation* (McGraw–Hill: New York, 1978). Another useful source of information is *Riegel's Handbook of Industrial Chemistry,* edited by James Kent (Van Nostrand Reinhold: New York, 1974).

Chapter 7

Two comprehensive and useful books on combustion technology have been published under the auspices of two of the leading companies in the field.

The books are *Combustion: Fossil Power Systems* by J. G. Singer (Combustion Engineering: Windsor, CT, 1981) and *Steam: Its Generation and Use* (Babcock and Wilcox: New York, 1978). A more recent book, which is particularly strong on fluidized-bed combustion, is David Merrick's *Coal Combustion and Conversion Technology* (Elsevier: Amsterdam, 1984). The effect of coal mineral matter in combustion systems is the subject of several chapters in *Mineral Matter and Ash in Coal,* edited by Karl S. Vorres (American Chemical Society: Washington, DC, 1986) and published in the ACS SYMPOSIUM SERIES.

Chapter 8

An old book, but one that still provides an excellent fundamental discussion of the manufacture of coke and its use in blast furnaces, is *Ferrous Metallurgy* (The Pennsylvania State College: State College, PA, 1939). Many middle-aged textbooks, which were used for high school or introductory college courses and published in the 1930s to early 1960s, before a systematic treatment of descriptive chemistry was excised from school curricula, contain extensive descriptions of coke, iron, and steel manufacturing. Often these books can still be found on library shelves; an example is *Chemistry in Modern Practice* by William Lemkin (Oxford Book Company: New York, 1943).

Chapter 9

Recovery of byproduct chemicals from carbonization is discussed in *Riegel's Handbook of Industrial Chemistry,* edited by James Kent (Van Nostrand Reinhold: New York, 1974). The events and people involved in establishing the coal-tar chemical industry are documented in Aaron Ihde's *The Development of Modern Chemistry* (Harper & Row: New York, 1964; recently published in paperback by Dover). Many of the books on chemistry or science written for lay readers and published prior to World War II contain extensive discussions of the discovery of coal-tar dyes and the subsequent growth of the industry. A classic of this genre is Edwin Slosson's *Creative Chemistry* (The Century Company: New York, 1919). Despite its age, Slosson's book still provides some enjoyable reading.

Chapter 10

Gasification technology is covered in *Riegel's Handbook of Industrial Chemistry,* edited by James Kent (Van Nostrand Reinhold: New York, 1974).

The topic of synthetic fuels is treated in *Synthetic Fuels* by Ronald Probstein and Edwin Hicks (McGraw-Hill: New York, 1981).

Chapter 11

In addition to the two books cited for Chapter 10, two books treat some aspects of coal liquefaction: Seymour Kaplan's *Energy Economics* (McGraw-Hill: New York, 1983) and Ruth Knowles's discussion of the energy crisis, *America's Oil Famine* (Coward, McCann and Geoghegan: New York, 1975). The role of catalysts in liquefaction and other processes is treated in an article entitled "Chemistry's Speedy Servants" by Lawrence Verbit, included in the 1982 *World Book Science Annual* (World Book: Chicago, 1982).

Index

A

B

C

D

E

F

G

H

I

M

N

O

Q

R

S

T

U

V

W

X

Editing by Karen L. McCeney
Indexing and production by Paula M. Bérard
Text design by Janet S. Dodd
Jacket design by Carla L. Clemens

Typesetting by Hot Type Ltd., Washington, DC
Jacket and covers printed by Atlantic Research Corporation, Alexandria, VA
Printed and bound by Maple Press, York, PA